Power, Testing, and Grounding of Electronic Systems

Power, Testing, and Grounding of Electronic Systems

Jesus C. de Sosa

iUniverse, Inc.
New York Bloomington

Power, Testing, and Grounding of Electronic Systems

iUniverse books may be ordered through booksellers or by contacting:

iUniverse
1663 Liberty Drive
Bloomington, IN 47403
www.iuniverse.com
1-800-Authors (1-800-288-4677)

ISBN: 978-0-595-47933-7 (sc)
ISBN: 978-1-4401-0725-2 (ebook)

Printed in the United States of America
iUniverse rev. date: 12/17/2008

Dedicated to

Anthony Jeric Diesta, Joseph Diesta, and Mia DeJesus

Contents

List of Tables

List of Figures

Preface

Power, Testing and Grounding of Electronic Systems addresses some of the critical issues in today's power distribution system and electronic system. Some of these issues include grounding noise current in a power distribution system, grounding noise voltage in an electronic system, redundant yet stable power system, and a testing methodology to systematically isolate the cause of a failure in a system.

These issues are costly. Incorrect grounding of facilities alone costs billions of dollars. Similarly, inadequate design of an electrical or electronic system may prevent a company from realizing its full market share. In the final analysis, a technical issue may become an economic issue; and, if left unresolved, may cause financial ruin.

In my experience, engineers, technicians, electricians, facility managers, and data center managers have found the information contained in this book useful. The construction, computer, telecommunications, power, and defense industries may also find benefits from the book.

Except for Laplace transform and a partial differential equation, the book used simple circuit analysis in deriving some of the equations in the book. Hence, no advanced mathematical knowledge is required. Where appropriate, the author developed figures to illustrate or clarify subtle points in a concept.

CHAPTER 1

Basics of Power Distribution System

This chapter presents the basic formulas for power calculations and the causes of high currents in the equipment grounding conductor. Formulas are represented as "per phase" instead of "single phase" power. This approach is simpler since it does not use the factor $\sqrt{3}$ as a multiplier for getting the line-to-line voltage from the line-to-neutral voltage. The practice is extensively used in advanced texts of power system analysis and design.

Modern high power digital loads create not only high unbalanced currents and but high grounding noise voltage as well. The former can affect the protective relays of a power distribution system. To provide a better understanding of how unbalanced currents are produced, circuit analyses of 4-wire and 5-wire 3-phase systems are made.

High grounding noise voltage can adversely affect the operation of sensitive electronic devices. Only a slight treatment of the issue is presented here. Later chapters on grounding will address the issue more comprehensively.

1.1   The Voltage, Phase, and Wiring of a Load in a Power Distribution System

Figure 1.1 shows the schematic diagram of a typical three phase power distribution system in a building. The figure consists of a transformer with delta primary winding and grounded-wye secondary winding. A three-phase load and a single-phase load are shown. The three-phase load has line-to-line voltage while the single-phase load has a line-to-neutral voltage. A delta to grounded-wye transformer is almost exclusively used in most facilities.

A transformer with a wye secondary winding may have the following types of loads:

1. three phase load with line-to-line voltages,
2. a single phase load with line-to-neutral voltage,
3. a single phase load with a line-to-line voltage, and
4. a mixture of three phase load and single phase loads.

A three-phase load has exactly four wires. Three of the wires are for the voltages. The remaining wire is the equipment grounding conductor. Engineers designate a three-phase load by specifying the voltage, the number of phases, and the number of wires. An example is "208-volt, 3-phase, 4-wire". The designation automatically means that three of wires are for line-to-line voltages and one of the wires in the 4-wire is an equipment grounding conductor.

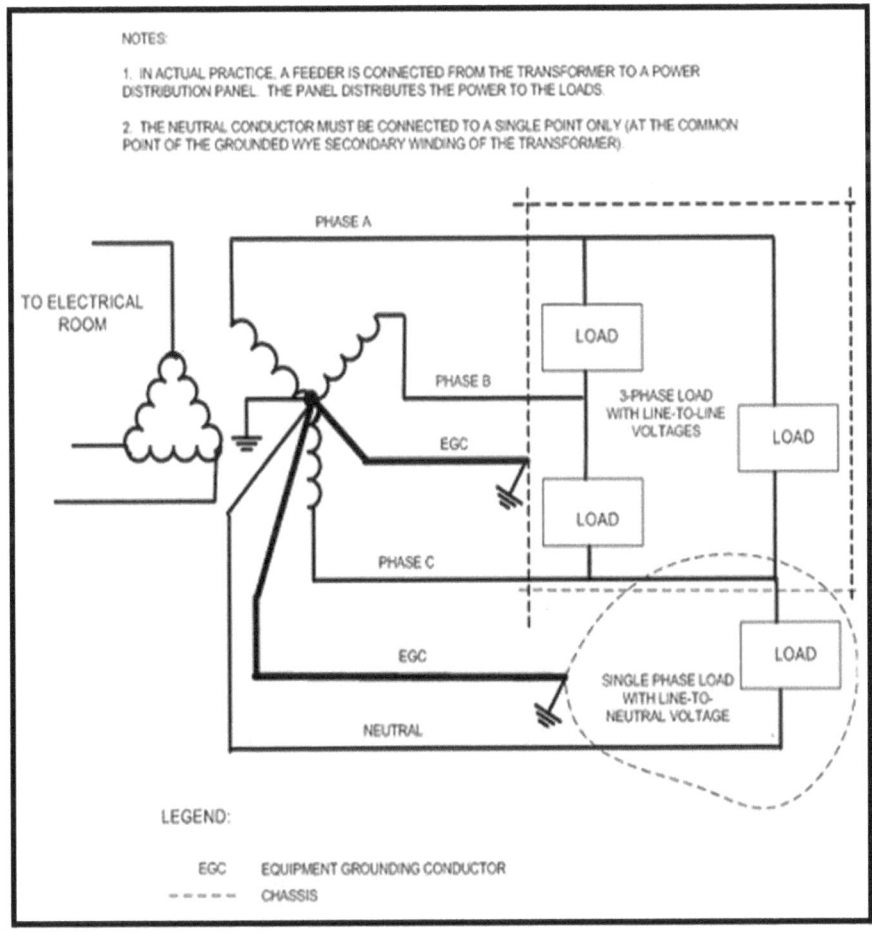

PHASE A

TO ELECTRICAL ROOM

LOAD

PHASE B

3-PHASE LOAD WITH LINE-TO-LINE VOLTAGES

LOAD

EGC

LOAD

PHASE C

EGC

LOAD

SINGLE PHASE LOAD WITH LINE-TO-NEUTRAL VOLTAGE

NEUTRAL

LEGEND:

EGC   EQUIPMENT GROUNDING CONDUCTOR
- - - - -   CHASSIS

**Figure 1.1 Representation of Grounded-Wye Secondary of
a Transformer with Delta-Connected Loads**

A single-phase load with line-to-neutral voltage has exactly three wires. One of the wires is a phase conductor, the second is the neutral, and the third is the equipment grounding conductor. An example of designation is "120-volt, single-phase, 3-wire". In this designation, two of the wires are for the line-to-neutral voltage and the remaining wire is for the equipment grounding conductor.

3

Single-phase load with line-to-line voltage has also three wires. Two of the wires are for the line-to-line voltage and the other is for the equipment grounding conductor. The designation follows the same convention above such as "208-volt, single-phase, 3-wire".

Sometimes, equipment may have a mixture of loads. That is, it has three phase loads and one or more single phase loads. Three possible cases are:

1. a mixture of three phase loads and single phase loads with line-to-neutral voltages,
2. a mixture of three phase loads and single phase loads with line-to-line voltages, and
3. a mixture of three phase loads and single-phase loads with line-to-line voltages and line-to-neutral voltage.

A slash indicates the mixture of loads and voltages. For the first case above, the designation is 208/120-volt, 3-phase, 5-wire. Three wires connect to the phases, the fourth wire is the neutral conductor, and the fifth wire is for the equipment grounding conductor.

The second case above has no line-to-neutral voltage. Therefore, its designation will be 208-volt, 3-phase, 4-wire. Again, three wires connect to the phases. The fourth wire is the equipment grounding conductor.

For the third case, the designation will be the same as the first case since it uses the three phases, the neutral conductor, and the equipment grounding conductor.

The examples above are taken from a 208-volt system. Another system commonly used is the 480-volt system. The same convention, to identify the voltage and wiring of the loads, applies.

Note that a transformer usually provides power to ten or more loads. To efficiently distribute the power and provide protection, a power distribution panel is installed between the transformer and the loads. The wires from the transformer to the panel are called feeder circuits, or simply feeder. Wires from the panel to the loads are the branch circuits.

A power distribution panel has a main circuit breaker and branch breakers. Inputs to the main breaker come from the phase conductors of the feeder. Inputs to the branch breakers come from the phase bars only. The main circuit breaker should trip in case a branch breaker fails to trip. It is another layer of protection.

The neutral of the load is connected to the neutral bar. Equipment grounding conductors are terminated in the grounding bar of the panel.

Figure 1.1 also shows the grounding of the neutral conductor. For a single-phase load with line-to-neutral voltage, the neutral conductor is the return wire. Whatever the current from the load is also flows in the neutral conductor. The neutral conductor of a single-phase load, with line-to-neutral voltage, must not be grounded with the equipment grounding terminal bar.

The neutral conductor of a feeder circuit carries the unbalanced current from a power distribution panel. It must be grounded at the common point of the grounded-wye secondary winding of a transformer.

The equipment grounding conductor is a safety conductor. It protects a person, touching the chassis of the equipment, when the chassis becomes "hot" or when a phase conductor inadvertently touches the chassis. Since the resistance of a person is much more than the resistance of the equipment grounding conductor, the parallel combination of the two is almost equal to the resistance of the equipment grounding conductor. As a result, the flow of power from the chassis will be diverted to the equipment grounding conductor and not the person.

An equipment grounding conductor is terminated with the equipment grounding bar of a power distribution panel. The bar must be shorted with the (metallic) chassis of the panel. If a metallic member of a facility is near the chassis, the chassis may also be grounded with the member. Unlike the neutral conductor, the equipment grounding conductor may be terminated at many points.

1.1.1   Formulas for Current and Voltage Regulation of a Three-Phase and Single-Phase Loads

A common problem in power design is sizing the wires to carry a known power of the load. The National Electrical Code provides information on the size of a wire given the current (of the load).

As described in the previous section, a load may be three-phase or a single-phase. The following formulas eliminate the use of the factor $\sqrt{3} = 1.732$. The factor converts a line-to-neutral voltage to line-to-line voltage.

Consider a three-phase load. The per phase power is given by

$$P_{1\phi} = P_{3\phi} / 3$$

where

$P_{1\phi}$ = per phase power, and

$P_{3\phi}$ = three phase power.

The current per phase (of a three-phase load) is

$$I_\phi = \frac{P_{1\phi}}{V_{LN}}$$

where,

I_ϕ = current per phase, and

V_{LN} = line-to-neutral voltage.

The above formula applies independent of whether a three-phase load has a neutral conductor or not.

For a single-phase load with line-to-line voltage, the current is

$$I_\phi = \frac{P_{1\phi}}{V_{LL}} .$$

The voltage drop across the ends of a wire feeding a load is

$$|\Delta V| = I_\phi Z_{wire}$$

where

ΔV = voltage drop across the wire, and

Z_{wire} = impedance of the wire.

The voltage regulation, R_V (in percent), of a wire with line-to-neutral voltage across a load is

$$R_V = \frac{|\Delta V|}{V_{LN}}(100\%).$$

If the load has line-to-line voltage, the voltage regulation of the wire is

$$R_V = \frac{|\Delta V|}{V_{LL}}(100\%).$$

Voltage regulation in power distribution system should be no more than 5.0%.

Note that the unit of power in the above formulas must be in volt-ampere, not watt. Additionally, all voltages must be the nominal system voltage of the source and not the load.

1.2   Increase of Current in the Equipment Grounding Conductor of a Transformer in a 3-Phase, 4-Wire System

There are cases when a three-phase load is connected to the secondary of a transformer without a neutral conductor. Such a load is usually designated as 208-volt or a 480-volt, 3-phase, 4-wire system.

The current in the equipment grounding conductor of a transformer feeding such a load may increase when the loads are not balanced. Since the current being measured is in the equipment grounding conductor, the current oftentimes is misrepresented as grounding current. In reality, it is an unbalanced current from the loads.

1.2.1　Case 1. Balanced Loads

Focus on node A of Figure 1.2, where the neutral terminal of a wye secondary (of a transformer), is grounded. Summing currents in the node gives the current, I_g, in the equipment grounding conductor as

$$I_g = I_a + I_b + I_c$$

Note:　Equipment grounding conductor of the load not shown for clarity.

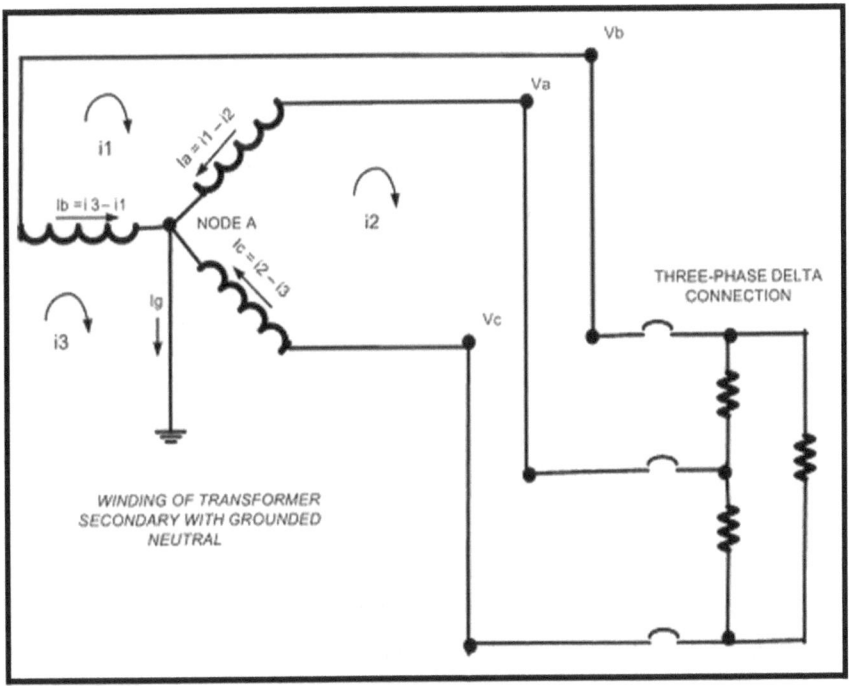

Figure 1.2　Currents in the Grounded Terminal of a Wye Secondary Winding

where

$$I_a = i_1 - i_2,$$
$$I_b = i_2 - i_3, \text{ and}$$
$$I_c = i_3 - i_1.$$

Note that the voltages producing the currents have different phase delays. For example, if the reference voltage V_a has zero phase, then V_b has 120 degrees phase shift from V_a; and, V_c has -120 degrees phase shift from V_a.

Consider a system with three phase loads. If the loads (represented as resistances) are equal and the voltages are equal then the loop currents are equal. That is,

$$i_1 = i_2 = i_3.$$

Therefore,

$$I_g = i_1 - i_2 + i_2 - i_3 + i_3 - i_1$$
$$= i_1 - i_1 + i_2 - i_2 + i_3 - i_3$$
$$= 0.$$

Because a term in the last equation is subtracted from the same term, the current in the equipment grounding conductor does not depend on the phase delay of any current, voltage, or load.

1.2.2   Case 2. Unbalanced Loads

Consider next that the loads or resistances are not equal. Assume for example that two of the resistances are open circuit and the other is one ohm. The resulting circuit is shown on Figure 1.3. For the loops with an open circuit, their corresponding loop current is zero. That is, the loop currents i_1 and i_2 are zero.

The current in the equipment grounding conductor is now

$$I_g = I_b + I_c.$$

Using the superposition theorem,

$$I_b = \frac{V_b}{R},$$

$$I_c = \frac{V_c}{R}, \text{ and}$$

$$I_g = \frac{V_b + V_c}{R}.$$

Voltages V_b and V_c differ from each other by their phase shift only. Both can be assumed to have one volt amplitude. That is, if

$$V_b = \cos(wt)$$

then

$$V_c = \cos(wt + 120).$$

Since $wt = 2\pi ft = 2\pi(t/T)$, where T is the period (equal to $1/f$, $f = 60$ Hz), values of voltages and currents can be computed as a function of t/T. Table 1 shows the results.

Since the assumed direction of the current, I_g, is downward, when the current on Table 1 is positive, then the unbalanced current is entering the node and the current in the conductor is leaving the node. Specifically, I_g leaves the node at 0%, 75%, and 100% of a cycle.

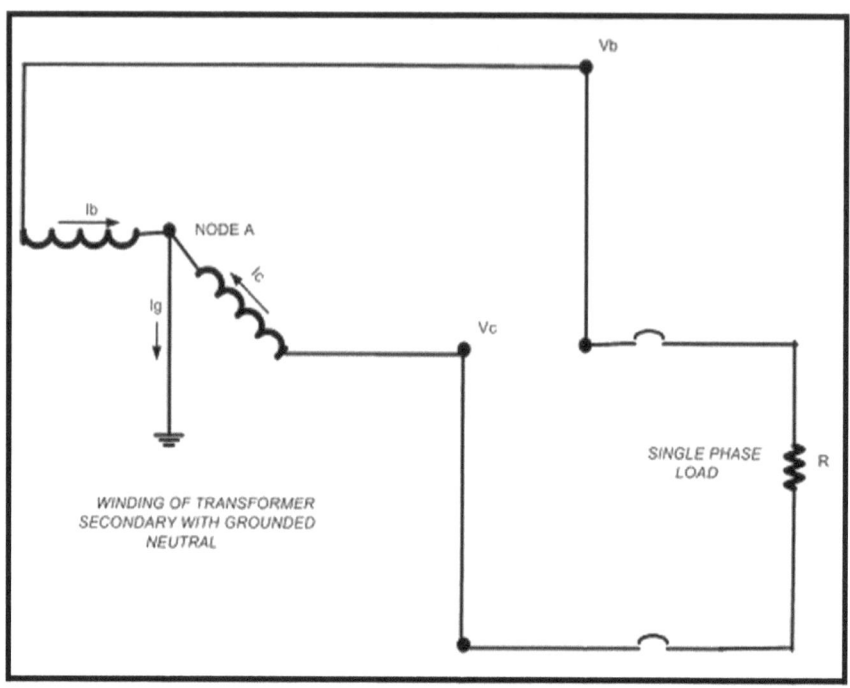

Figure 1.3 Currents when Two of the Loads in a Transformer are Open Circuit

Table 1.1 Normalized Values of Instantaneous and RMS Currents with a Load of One Ohm

t/T	Ib	Ic	Ig	Remark
0 (0% of a cycle)	1.00	-0.50	0.50	Ground current leaving the node
¼ (25% of a cycle)	0.00	-0.86	-0.86	Ground current entering the node
½ (50% of a cycle)	-1.00	0.50	-0.50	Ground current entering the node
¾ (75% of a cycle)	0.00	0.86	0.86	Ground current leaving the node
1 (100% of a cycle)	1.00	-0.50	0.50	Ground current leaving the node
RMS (approx.)	0.774	0.664	0.664	

Conversely, when the neutral current is negative then the unbalanced current is leaving the node and the current in the equipment grounding conductor is entering the node. This condition occurs at 25% and

50% of a cycle. Since the conductor is connected to the earth, the earth itself supplies the current to the node.

Note that at any time, the instantaneous neutral current can be positive (entering the node), or negative (leaving the node). The RMS values are calculated for five samples only. As the number of samples increases, the RMS values should approach $\left|\sqrt{2}\right|/2$.

1.3 Increase of Current in the Equipment Grounding Conductor of a Transformer in a 3-Phase, 5-Wire System

A 3-phase, 5-wire power distribution system has a mixture of loads with line-to-line voltages, and line-to-neutral voltage. It is designed to provide power not only to 3-phase loads but to single-phase loads as well.

If some loads in the system have line-to-line voltages, then the loads have no neutral current or conductor. They, however, have equipment grounding conductors. The unbalanced currents caused by such loads flow thru the equipment grounding conductor of a grounded-wye secondary via the phase windings of the transformer. Its mechanism is the same as when the loads are fed by a 3-phase, 4-wire system (see previous section).

If some of the loads in the 3-pahse, 5-wire system are single-phase loads (with line-to-neutral voltages), then the unbalanced currents from such loads flow through the neutral feeder and finally to the equipment grounding conductor (of the transformer).

1.4   High Frequency Grounding Noise
Voltage in Electronic Equipment

Consider a room with arrays of electronic loads such as racks of computers in a data center or racks of automated test equipment in a test facility. Such loads have high-power switching components that convert the AC voltage to DC. Switching signals are

in the form of a pulse width modulated signals or pulse amplitude signals. Independent of their forms, the waveform generates high-power high-frequency harmonics. Typical frequency range of the harmonics ranges from a few kilohertz to a few megahertz.

A method to measure the intensity of such harmonics is by measuring the voltage across the chassis of the electronic load and a facility steel (or any other metal) member. If the chassis-to-facility measurement reads unusually high root-mean-square voltages, then there is a lot of switching happening in the equipment.

A 1.12-volt RMS reading, for example, may mean that the equipment produces zero-to-peak transients of 20 to 25 volts in the frequency range of 5 Hz to 10 KHz. The pulse widths of the transients may range from several nanoseconds to a few microseconds. Because the pulse widths are narrow, the energy of the pulse is sufficiently small so as not to cause a safety hazard. The pulses, however, may impair the performance of the loads.

Like any other equipment, electronic loads have equipment grounding conductor. Such a conductor, oftentimes, may not be sufficient to get rid of the harmonics. As a result, each electronic load becomes a transmitter and receiver of noise. Furthermore, the noise voltages produce currents in the equipment grounding conductor of a transformer. Since current in the equipment grounding conductor of the transformer flows from the power system to the earth and vice-versa, the transformer itself becomes a receiver and transmitter of noise. In chapter 8, the design of well grounded facility with electronic loads will be proposed.

CHAPTER 2

Calculations of Short Circuit or Fault Currents

Equipment must be protected against electrical fault or short circuit. An electrical fault may cause the loss of the equipment, its facilities and in some cases lives. The first step in such a protection is the calculation of the fault currents at the points of faults.

All circuit breakers, whether electro-mechanical or electronic, have instantaneous (or magnetic) and thermal tripping mechanisms. During a fault (short circuit), the instantaneous (or magnetic) tripping mechanism trips a circuit breaker. Otherwise, during an overload the thermal tripping mechanism trips the circuit breaker.

A circuit breaker must have sufficient short circuit interrupting capacity. Circuit breakers of the 600-volt class and 100-ampere frame have minimum interrupting capacity of 10,000 amperes. Other ampere frames can have as much as 65,000 amperes of interrupting capacity. Fault calculations help ensure that a selected circuit breaker has enough interrupting capacity. Otherwise, fusing or melting of contacts, may occur.

2.1   Fault Calculations at the Output of Electronic Equipment

Figure 2.1 shows equipment with facility power input, four 12 VDC DC power supplies with their own over current protection, and loads. A 2-pole circuit breaker protects the equipment. The breaker is rated at 208 volts.

The intensity of the fault currents depends on the resistance of the shorting switch. A switch with higher resistance may see lower levels of faults than a switch with lower resistance. Low level of fault currents, however, may fail to trip the circuit breaker. Calculations of fault currents must, therefore, show if the circuit breaker provides adequate protection.

Three examples of fault calculations will be shown. The first is on the output side of the equipment with a fault resistance that is relatively large. A second calculation is on the output side also but with relatively small resistance. The last calculation is on the input side of the equipment immediately after the output of the circuit breaker.

Figure 2.1 shows the single line diagram of the equipment with a fault on the 12 VDC. output. The figure shows also the actual measured values of the parallel resistance and capacitance across the output. Two cases of hypothetical fault resistance are shown: (1) 0.000645 ohm, and (2) 0.0001 ohm.

Figure 2.2 shows the representation of the resistances and the capacitance. The two parallel resistances must be combined to a single resistance. After the conversion, the resulting resistance in parallel with the capacitance must be converted to equivalent resistance and equivalent inductance in series. The parallel-to-series conversion is required since the formula for fault current uses inductance as a parameter. Figure 2.3 shows the equivalent resistance in series with the equivalent inductance.

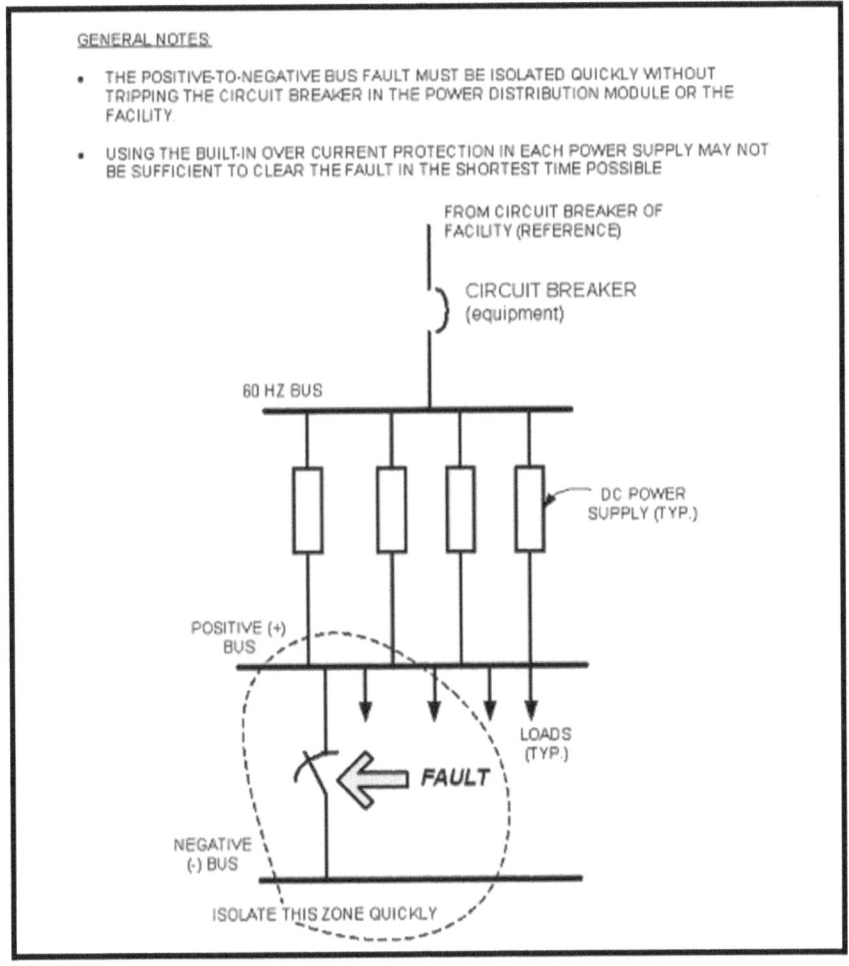

**Figure 2.1 Single Line Diagram of the Equipment with a
Fault across the Positive and Negative Busses**

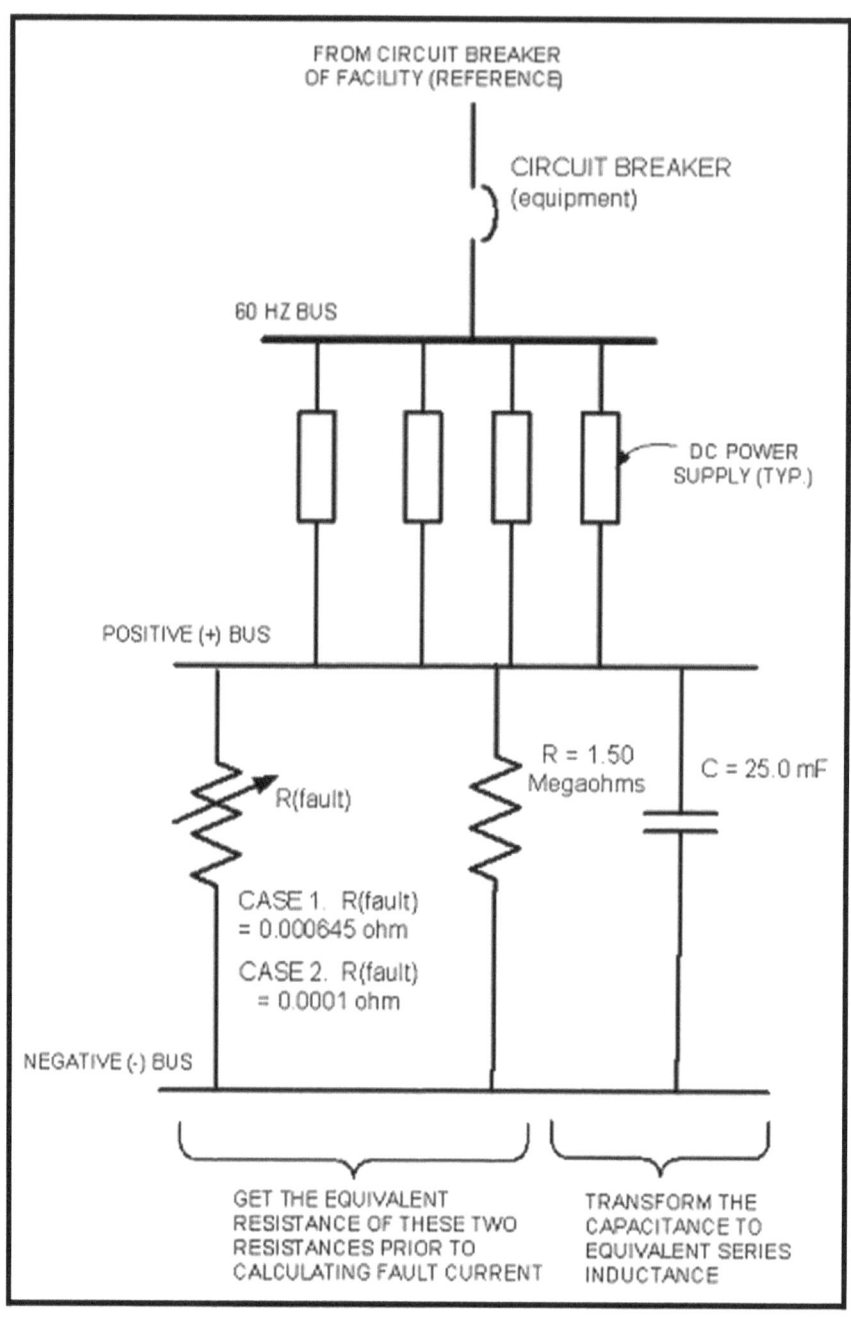

Figure 2.2 Assumed Fault Resistance and Measured Resistance and Capacitance Prior to a Fault

Use the following formulas to perform the parallel-to-series conversion:

$$R_{equiv} = \frac{R}{1+(\omega RC)^2}$$

and

$$L_{equiv} = -\frac{\omega R^2 C}{1+(\omega RC)^2}$$

where

ω = frequency,

R = resistance in parallel,

C = capacitance in parallel,

R_{equiv} = equivalent resistance in series, and

L_{equiv} = equivalent inductance in series.

The following equation will be used in all fault calculations:

$$i(t) = \frac{V_{max}}{Z}\left[\sin(\omega t + \alpha - \theta) - e^{-Rt/L}\sin(\alpha - \theta)\right]$$

where,

t = time,

α = electrical angle after the switching of the fault,

θ = power angle, and

Z = impedance across the fault.

The worst-case electrical angle (after the switching of the fault) occurs at around 40 to 70 degrees corresponding to about 0.69 to 1.22 radians. At these angles the voltage waveform is getting close to its peak value. Power angle is angle between the reactance of the equivalent inductance and the equivalent resistance. Z is the magnitude of the reactance of the equivalent inductance and the equivalent resistance.

Since the above formulas will require repetitive calculations as a function of time, a computer program may be developed to automate the calculations. Appendix A shows a sample program written in Java programming language.

Table 2.1 and Table 2.2 show the results of calculations at the output side of the equipment. The fault current at the output side must be converted to the input side. For example, Table 2.1 shows the average fault current for all data is approximately equal to 4,948.0 amperes (at 12 VDC bus). This is equivalent to 59,376.0 watts. At the 60 Hz input side and with 208-volt line-to-line (single-phase) voltage, the fault current is 286.0 amperes.

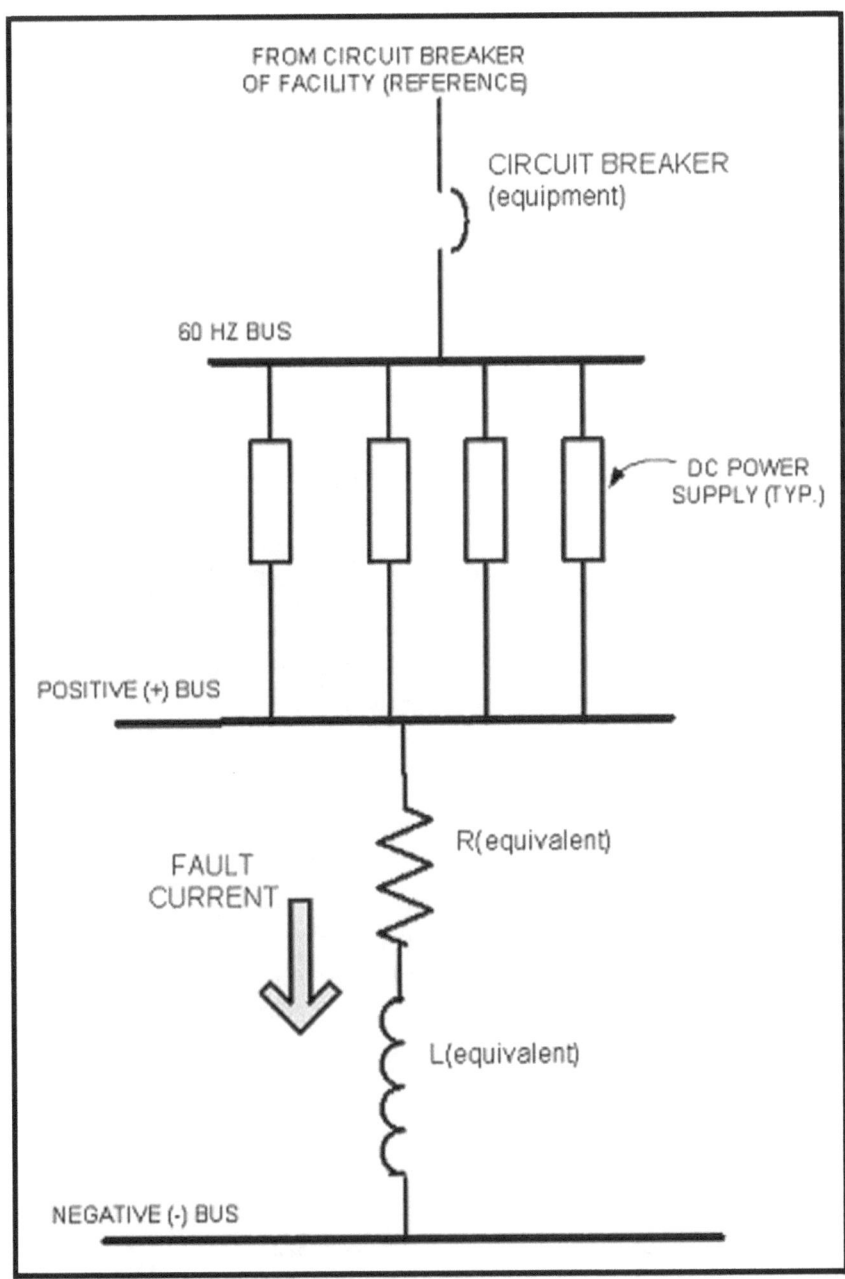

**Figure 2.3 Equivalent Resistance and Equivalent
Inductance of the Circuit from Figure 2.2**

**Table 2.1   Result of Short Circuit Simulation using a Shorting
Switch with 0.000645 Ohm (Equivalent to a 0.50 foot #10
AWG Conductor)**

TIME (second)	CURRENT (amperes)
0.0	0.0
6249.999	4948.079185008196
12499.999	4948.079184925934
18749.999	4948.079184420686
24999.999	4948.07918476141
31249.999	4948.079184256162
37499.999	4948.079183750913
43749.999	4948.079184373628
49999.999	4948.07918443236
56249.999	4948.079184491094
62499.999	4948.079183421864
68749.999	4948.079183480597
74999.999	4948.079182411368
81249.999	4948.079182470101
87499.999	4948.079183656797
93749.999	4948.079182587567
99999.999	4948.079183774263
106249.999	4948.079182705033
112499.999	4948.0791838917285
118749.999	4948.079182822499
124999.999	4948.07918175327
131249.999	4948.07918068404
137499.999	4948.079181870736
143749.999	4948.0791808015065
149999.999	4948.079179732277
156249.999	4948.079180918973
162499.999	4948.079179849743
168749.999	4948.079178780514
174999.999	4948.079182223135
181249.999	4948.079181153905
187499.999	4948.0791800846755
193749.999	4948.079179015446

Table 2.2   Result of Short Circuit Simulation using a Shorting Switch with 0.0001 Ohm

TIME (second)	CURRENT (amperes)
0.0	0.0
62500.0	29688.47510729896
125000.0	29688.475104055167
187500.0	29688.475107579154
250000.0	29688.47509756759
312500.0	29688.475114627123
375000.0	29688.475104615554
437500.0	29688.47509460399
500000.0	29688.475084592425
562500.0	29688.475074580856
625000.0	29688.475118711493
687500.0	29688.475108699924
750000.0	29688.47509868836
812500.0	29688.475088676794
875000.0	29688.475078665226
937500.0	29688.47506865366
1000000.0	29688.475058642092
1062500.0	29688.475048630527
1125000.0	29688.475038618963
1187500.0	29688.475028607394
1250000.0	29688.475126880232
1312500.0	29688.475116868663
1375000.0	29688.4751068571
1437500.0	29688.47498856113
1500000.0	29688.475086833965
1562500.0	29688.474968537997
1625000.0	29688.47506681083
1687500.0	29688.475165083666
1750000.0	29688.475046787702
1812500.0	29688.475145060533
1875000.0	29688.47502676457
1937500.0	29688.475125037403

Table 2.2, however, shows larger fault current. It is about 29,688.0 amperes (at 12 VDC bus). This is equivalent to about 356,256.0 watts. At the 60 Hz input side and 208-volt line-to-line (single-phase) voltage, the interrupting capacity is 1,713.0 amperes.

2.2   Fault Calculation at the Input of Electronic Equipment

The same procedure is also employed when calculating the fault current at the input side. However, an additional step is required. That is, the resistance and inductance of the wires feeding the equipment must also be taken into account.

Figures 2.4 through 2.6 show the steps. Note that on Figure 2.6, the equivalent resistance of the equipment must be added to the resistance of the wire (feeding the equipment). Similarly, the equivalent inductance must be added to the inductance of the wire.

Table 2.3 shows the calculations at the input side. Unlike Tables 2.1 or 2.2, the fault current exponentially increases. After one millisecond, the fault current reaches 4,979.40 amperes. After one cycle of the 60-Hz voltage waveform, the fault current increased to 5,527.18 amperes. The increase is about 11.0% of the initial value. In two cycles, the fault current jumps to 6,007.98 amperes – a 21.0% increase over the initial value at one millisecond.

The above calculations show that the minimum interrupting capacity of the circuit breaker must be at least 6,007.98 amperes. A circuit breaker rated with 10,000-ampere interrupting capacity might just as well be used.

2.4   A Note on the Line to Ground Fault of a Load
Connected from the Delta Winding of a Transformer

Figure 2.7 is the schematic diagram of a wye-to-delta transformer.
Loads are connected across the delta windings. If there is a line-to-
ground fault in one of the windings, the voltage does not go down to
near zero but between the line-to-neutral voltage of the wye winding
and the line-to-line voltage of the wye winding. Circuit breakers
designed for such applications must be carefully selected.

The above discussion does not hold true if the loads are connected
across the wye windings although the loads may not have a neutral
conductor. In such a case, a line-to-ground fault will tend to bring
down the voltage near zero.

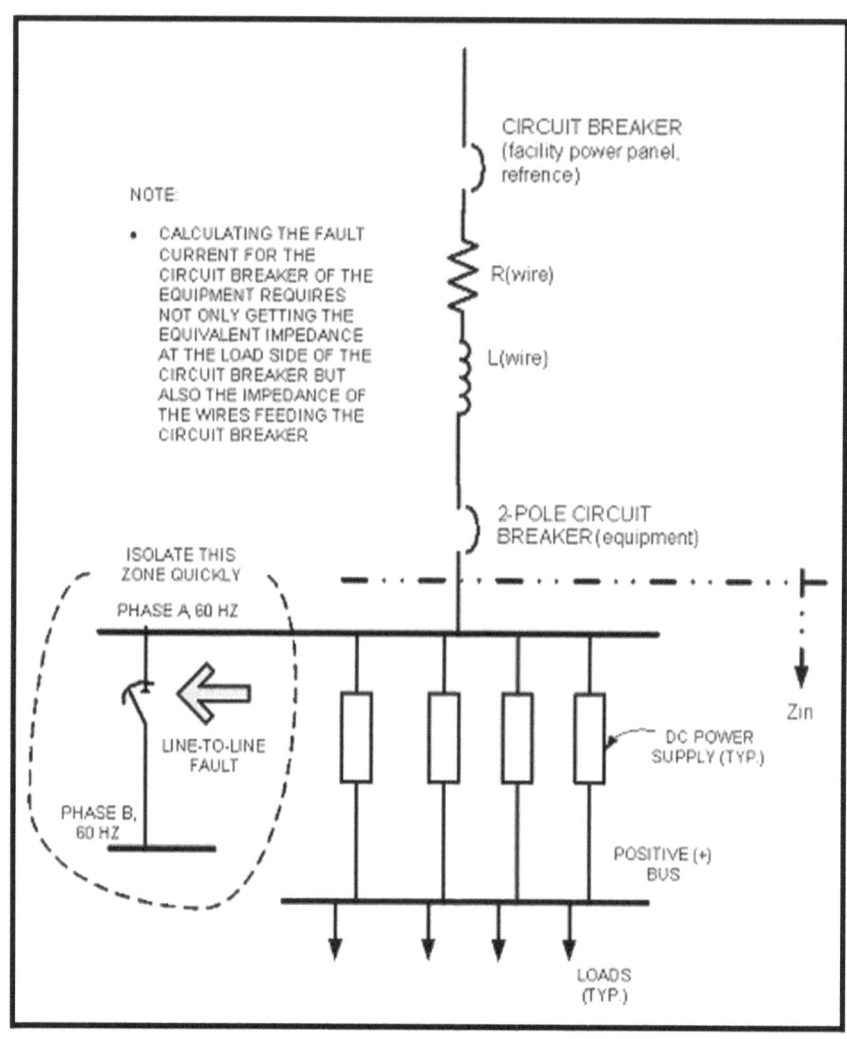

CIRCUIT BREAKER
(facility power panel, refrence)

NOTE:

- CALCULATING THE FAULT CURRENT FOR THE CIRCUIT BREAKER OF THE EQUIPMENT REQUIRES NOT ONLY GETTING THE EQUIVALENT IMPEDANCE AT THE LOAD SIDE OF THE CIRCUIT BREAKER BUT ALSO THE IMPEDANCE OF THE WIRES FEEDING THE CIRCUIT BREAKER

R(wire)

L(wire)

2-POLE CIRCUIT BREAKER (equipment)

ISOLATE THIS ZONE QUICKLY

PHASE A, 60 HZ

LINE-TO-LINE FAULT

PHASE B, 60 HZ

Zin

DC POWER SUPPLY (TYP.)

POSITIVE (+) BUS

LOADS (TYP.)

Figure 2.4 Single Line Diagram of the Equipment with a Fault on the Input Side

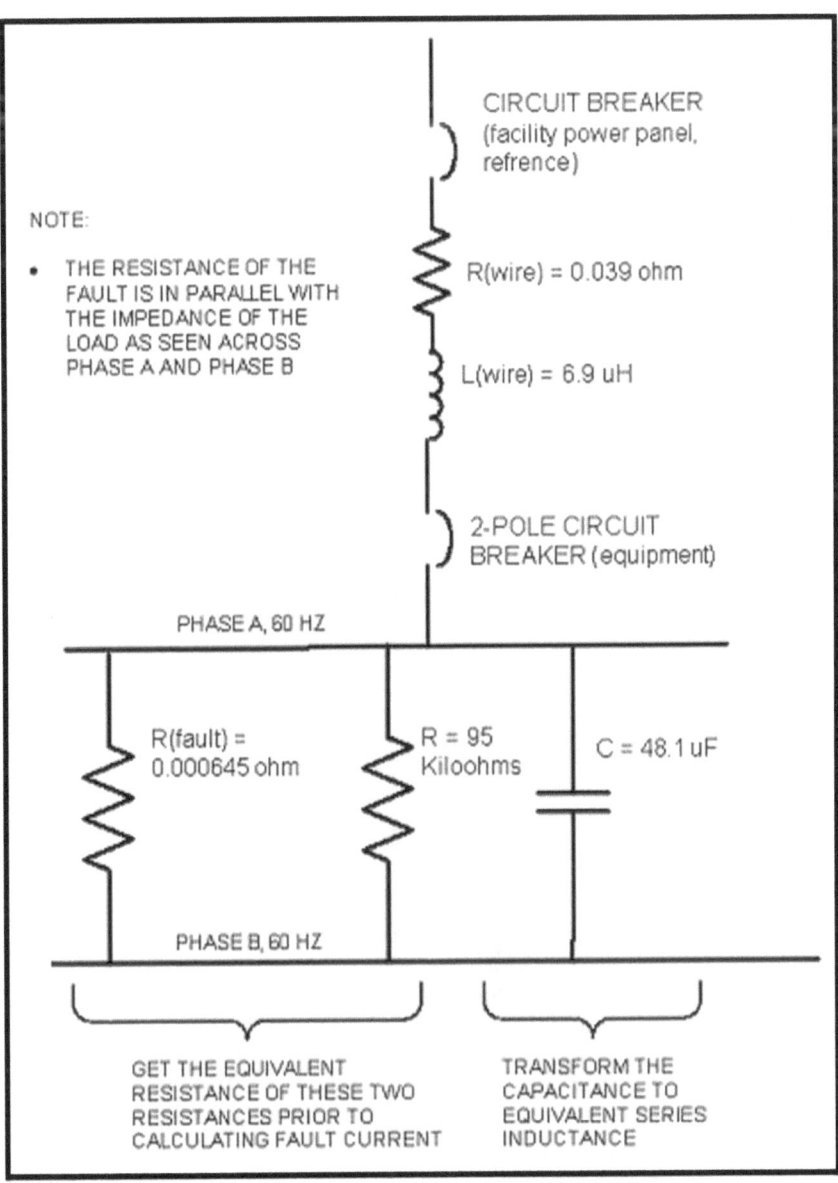

Figure 2.5 Circuit Representation of the Fault Resistance, and the Measured Resistance and Capacitance prior to the Occurrence of the Fault

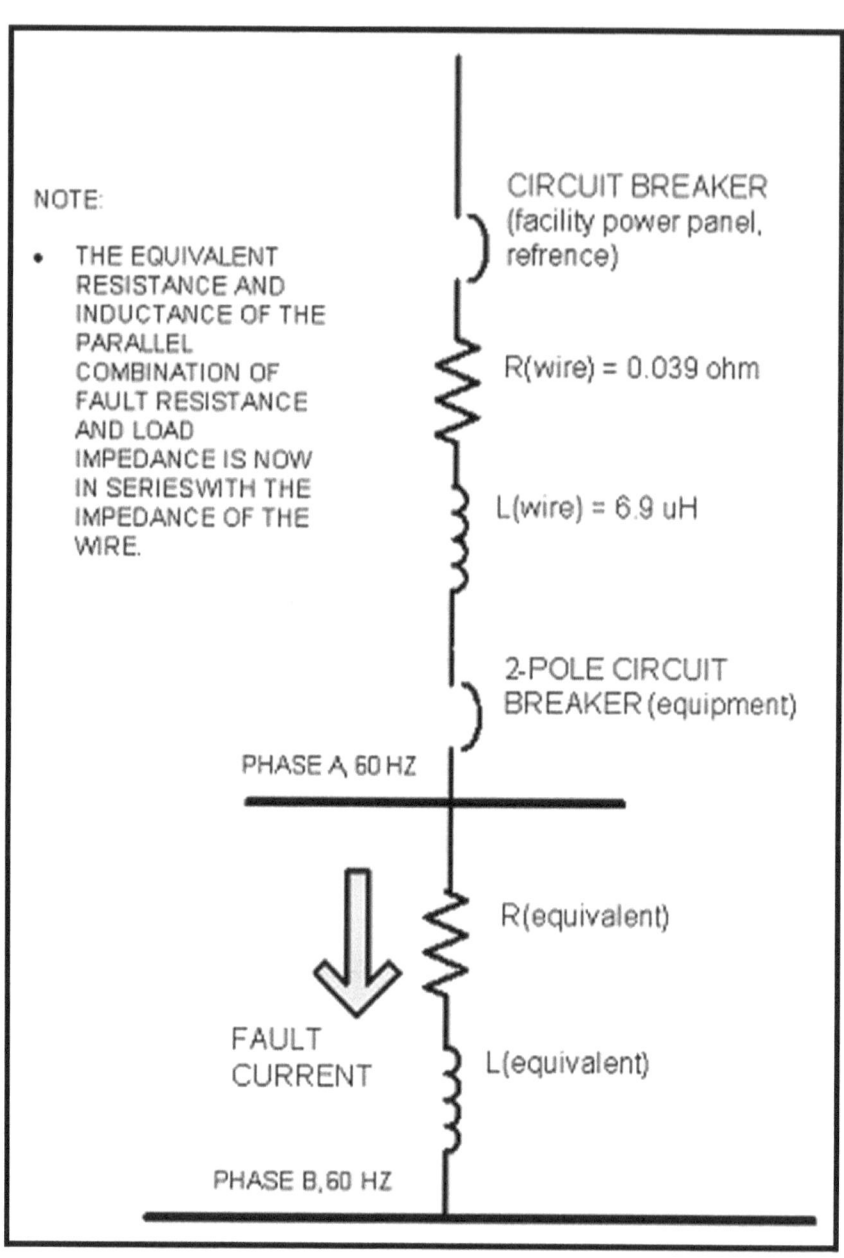

Figure 2.6 Further Reduction of the Circuit from the Figure 2.5

Table 2.3   Computer Simulation of Line-to-Line Electrical Fault across the Phases of a Circuit Breaker whose Circuit Parameters are shown on Figure 2.6

TIME (second)	CURRENT (amperes)
0.0	0.0
0.0010	4979.409371804139 - (in1 millisecond)
0.0020	5030.273173554674
0.0030	5067.253256855501
0.0040	5103.978368395493
0.0050	5140.484737356879
0.0060	5176.7709047694025
0.0070	5212.83531654519
0.0080	5248.676427805675
0.0090	5284.292703236946
0.01	5319.6826171563025
0.011	5354.8446535776175
0.012	5389.77730627628
0.013	5424.479078853713
0.014	5458.948484801475
0.015	5493.1840475649415
0.016	5527.184300606545 - (in 16 milliseconds or1 cycle of 60 Hz)
0.017	5560.947787468609
0.018	5594.473061835728
0.019	5627.758687596731
0.02	5660.803238906191
0.021	5693.605300245514
0.022	5726.163466483562
0.023	5758.4763429368595
0.025	5790.542545429322
0.026	5822.360700351561
0.027	5853.929444719717
0.028	5885.247426233851
0.029	5916.31330333587
0.03	5947.125745266993
0.031	5977.683432124761
0.032	6007.985054919571 - (in 32 milliseconds or 2 cycles of 60 Hz)

Note: The above data agrees well with the data on Table 7-8 – Single-pole short circuit test values for MCCBs (molded case circuit breaker) of the IEEE Std 242-2001. Table 7-8 shows that for a frame rating of 100 amperes and 250-volt maximum, the RMS symmetrical fault current of a two-pole breaker is 5,000 amperes.

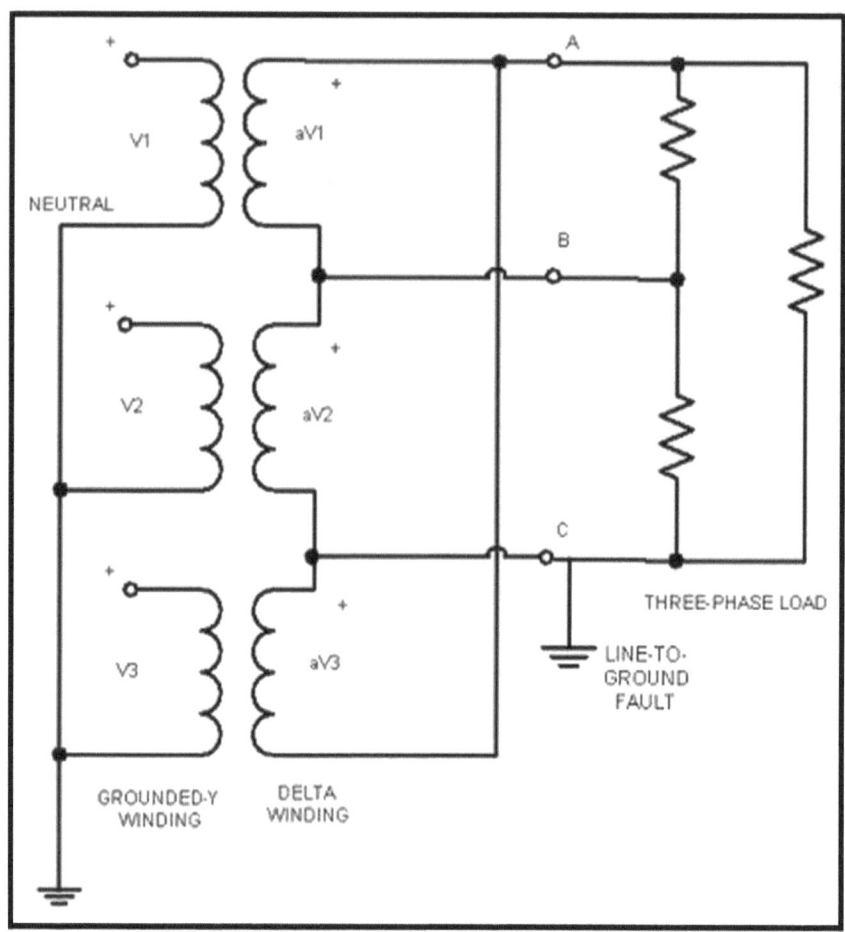

Figure 2.7 Line-to-Ground Faults in a Grounded Y to Delta Transformer

CHAPTER 3

Propagation Constant and Characteristic Impedance of Transmission Line

A transmission line is any pair of wires or other conducting media that transmits power from a source to a load. Understanding a transmission line is fundamental in electrical engineering since a line may attenuate or delay a signal. Similarly, the impedance of a line determines the maximum power that can be transferred from the source to the load.

3.1   Propagation Constant of a Transmission Line

Consider any transmission line. Its propagation constant given by

$$\gamma = \alpha + j\beta$$

where,

γ = propagation constant,

α = attenuation constant in Nepers/meter, and

β = phase constant in radians/meter.

Note that the real part of the propagation constant is the attenuation constant while its imaginary part is the phase constant.

As shown on Figure 3.1, an increment of a transmission line has the impedance

$$Z = R + j\omega L$$

Its corresponding shunt admittance is

$$Y = G + j\omega C$$

G is the conductance (or real part) of the admittance and wC is the susceptance (or the imaginary part) of the admittance. Note that all parameters are in per unit length.

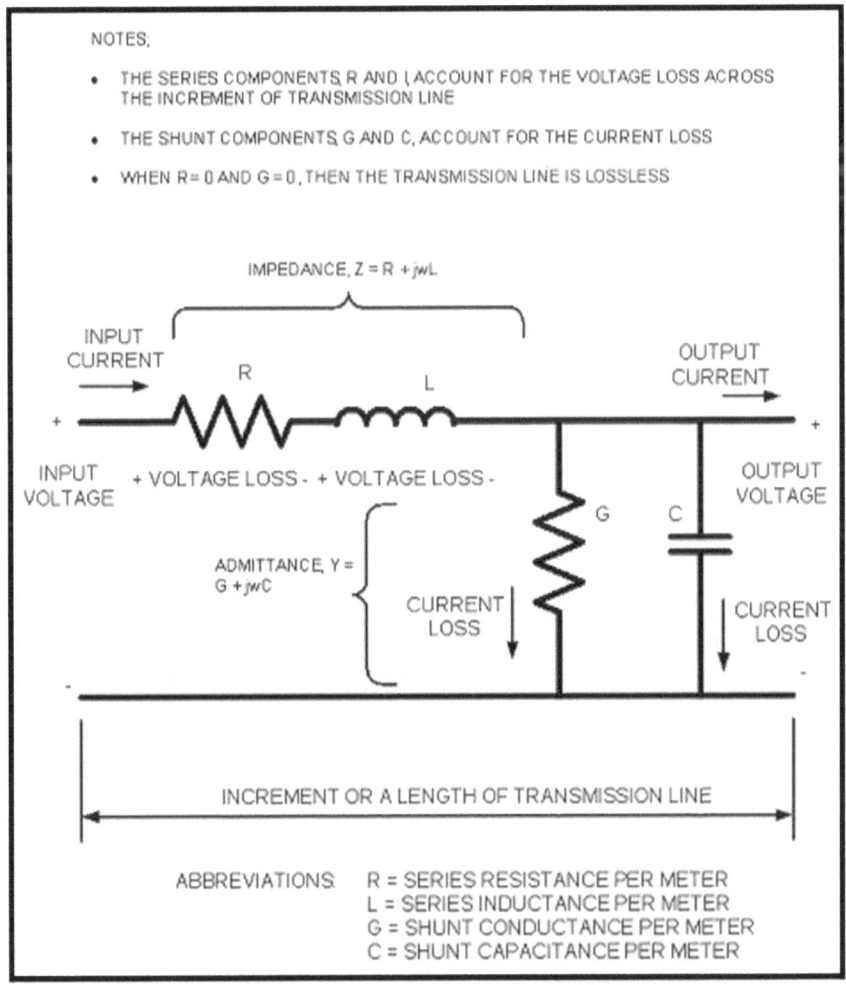

Figure 3.1 The Basic Elements of a Transmission Line

Using Kirchoff's voltage law, the partial derivative of the voltage across the line is

$$\frac{\partial V(x)}{\partial x} = -(R + jwL)I(x)$$

Similarly, the partial derivative of current is

$$\frac{\partial I(x)}{\partial x} = -(G + jwC)V(x)$$

Taking the partial derivative again gives

$$\frac{\partial V^2(x)}{\partial x^2} = \gamma^2 V(x), \text{ and}$$

$$\frac{\partial I^2(x)}{\partial x^2} = \gamma^2 I(x)$$

where

$$\gamma = \sqrt{ZY} = \sqrt{(R + j\omega L)(G + j\omega C)}$$

As a function of distance, x, the voltage and current are given by:

$$V(x) = V\_e^{-\gamma x} + V_+ e^{+\gamma x},$$

and

$$I(x) = I\_e^{-\gamma x} + I_+ e^{+\gamma x}$$

The above formulation of the partial differential equation of a transmission line and its solution are so important that they are repeated on Figures 3.2 and Figure 3.3.

Note that the propagation constant is the square root of a complex number. Its computation may be impossible unless it could be reduced to a simpler form.

Let

$$\gamma = \sqrt{a + jb}$$

where,

$$a = RG - \omega^2 LC \text{, and}$$
$$b = \omega GL + \omega RC$$

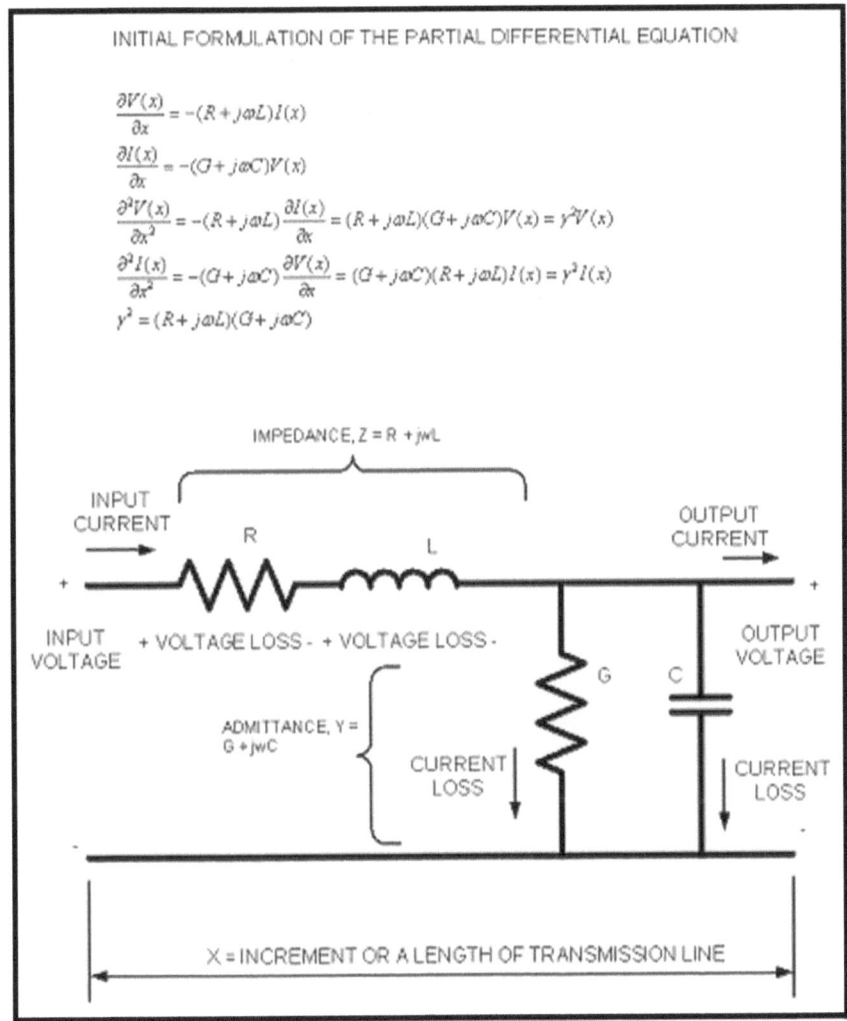

Figure 3.2 **The Transmission Line with the Initial Formulation of its Partial Differential Equation**

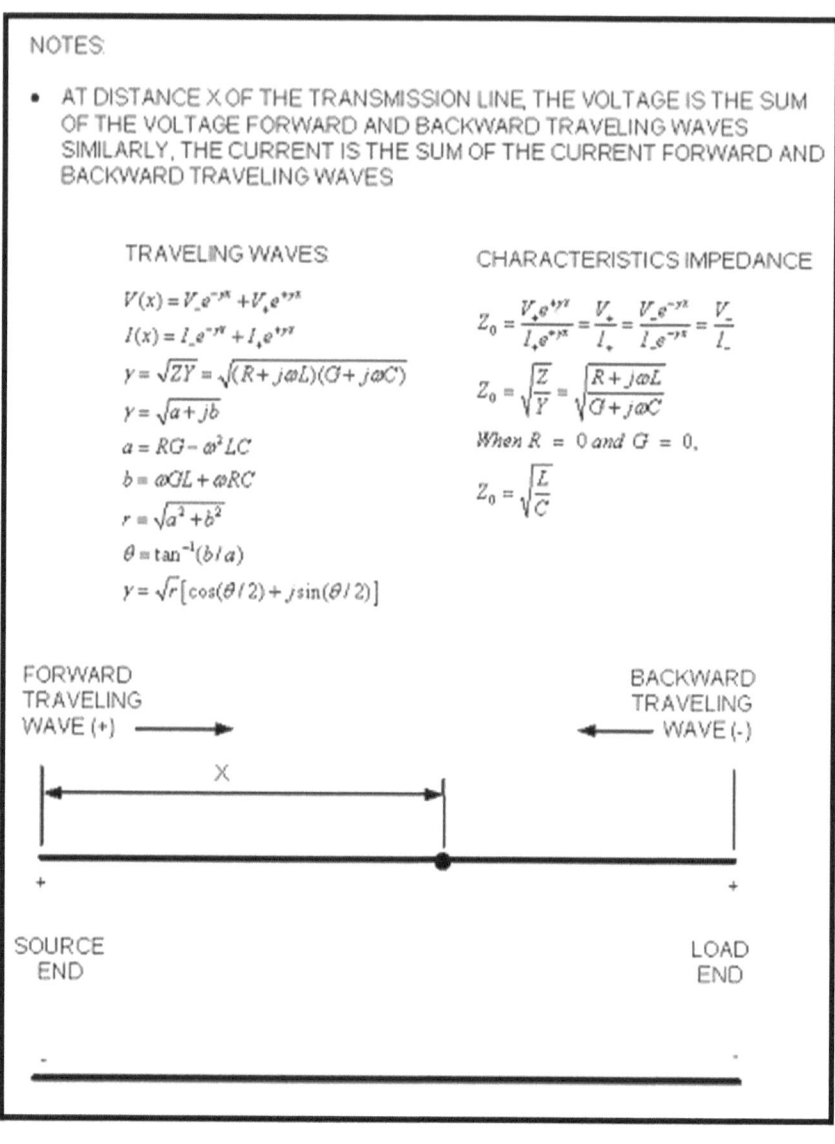

Figure 3.3 A Transmission Line has Forward and Backward Traveling Waves

Using DeMoivre's formula,

$$\gamma = \sqrt{r}\left(\cos(\theta/2)\right) + j\sqrt{r}\left(\sin(\theta/2)\right)$$

where,

$$r = \sqrt{a^2 + b^2}, \text{ and}$$

$$\theta = \tan^{-1}(b/a)$$

Since $\gamma = \alpha + j\beta$, the real (or attenuation) component of the propagation constant is

$$\alpha = \sqrt{r}\cos(\theta/2)$$

Its imaginary (or phase constant) component is

$$\beta = \sqrt{r}\sin(\theta/2)$$

The two constants differ only by the cosine and sine of the angle between the imaginary and real components of the product of the series impedance and shunt admittance.

The special case when $R = 0$ and $G = 0$ gives

$$\alpha + j\beta = \sqrt{-\omega^2 LC} = 0 + j\omega\sqrt{LC}$$

so that

$$\alpha = 0, \text{ and}$$

$$\beta = \omega\sqrt{LC}$$

When $R = 0$ and $G = 0$, there is no power loss, or attenuation of signal, in the line.

Since $\beta = \dfrac{2\pi}{\lambda}$, the wavelength, λ, of the signal is

$$\lambda = \frac{1}{f\sqrt{LC}}$$

The factor, $1/\sqrt{LC}$, is called the phase velocity (approximately equal to the speed of light). In chapter 8, the wavelength will be used as a parameter in designing a grounding system.

3.2   Characteristic impedance of a lossless line

In general, the characteristics impedance, Z_0, is the ratio of the impedance and admittance of the line,

$$Z_0 = \sqrt{\frac{R + j\omega L}{G + jwC}}$$

When the line is lossless ($R = 0$ and $G = 0$), the attenuation constant is zero and the characteristic impedance takes the form

$$Z_0 = \sqrt{\frac{L}{C}}$$

The unit of the characteristic impedance is ohm although it is not the conventional resistance, R. It is simply the impedance of a line with no series resistance or shunt conductance.

An estimate of the characteristic impedance can be made by using

$$Z_0 = \sqrt{Z_{SC} Z_{OC}}$$

where

Z_0 = charactereistics impedance,

Z_{SC} = impedance when the load end is short circuited, and

Z_{OC} = impedance when the load end is open circuited.

Impedance matching, or conjugate impedance matching, is based on designing a circuit to match the characteristic impedance of a line. It is especially important when the physical length of the line is comparable to the wavelength of the signal. This condition occurs on RF circuits (where wavelengths are small) and on the transmission of the 60 Hz electrical power across several hundreds or thousands of kilometers.

3.2.1 Characteristics impedance and impedance matching of a lossy line

If the series resistance, R, and the shunt conductance, G, are not zero then the characteristics impedance is the original formula

$$Z_0 = \sqrt{\frac{R + j\omega L}{G + jwC}} = A + jB$$

If the load impedance is

$$Z = R + jX$$

then the real part of the line must be equal to the resistive part of the load. The imaginary part of the line, however, must be the complex conjugate of the imaginary part of the load. Symbolically,

$$A = R,$$

and

$$jB = -jX$$

The DeMoivre's formula may be used to evaluate A and B in a manner similar to finding α and β of the propagation constant.

3.2.2   Surge Impedance Loading in Power Transmission

The power industry defines a figure of merit called surge impedance loading (SIL) to assess the stability of power transmission in long lines.

It is defined as

$$SIL = \frac{V_{LL}^2}{Z_0}$$

Since Z_0 is in ohms, then SIL is in unit of watts. Note that the formula assumes a line-to-line voltage. Similar formula for line-to-neutral voltage may be used provided the characteristic impedance is measured relative to the neutral line.

The SIL also provides information on the thermal limit, voltage regulation limit, and stability limit of a transmission line. Figure 3.4 shows these limits.

To identify critical points in the graph of a typical *SIL*, an exponential function fits the graph. Argument of this function is the normalized distance, x. The normalizing distance, L, is the distance that corresponds to the ideal case when the ratio of the actual loading and *SIL* is 1.0.

At distances far away from a generator, a load sees less power. As a result, the equivalent power delivered by a generator to a load is less. When such a condition occurs, instability occurs.

Conversely, as the distance gets closer to the generator, the power delivered by the generator is more than the optimum value. For example, at $x = 2$, the ratio of the actual loading and *SIL* is 1.414.

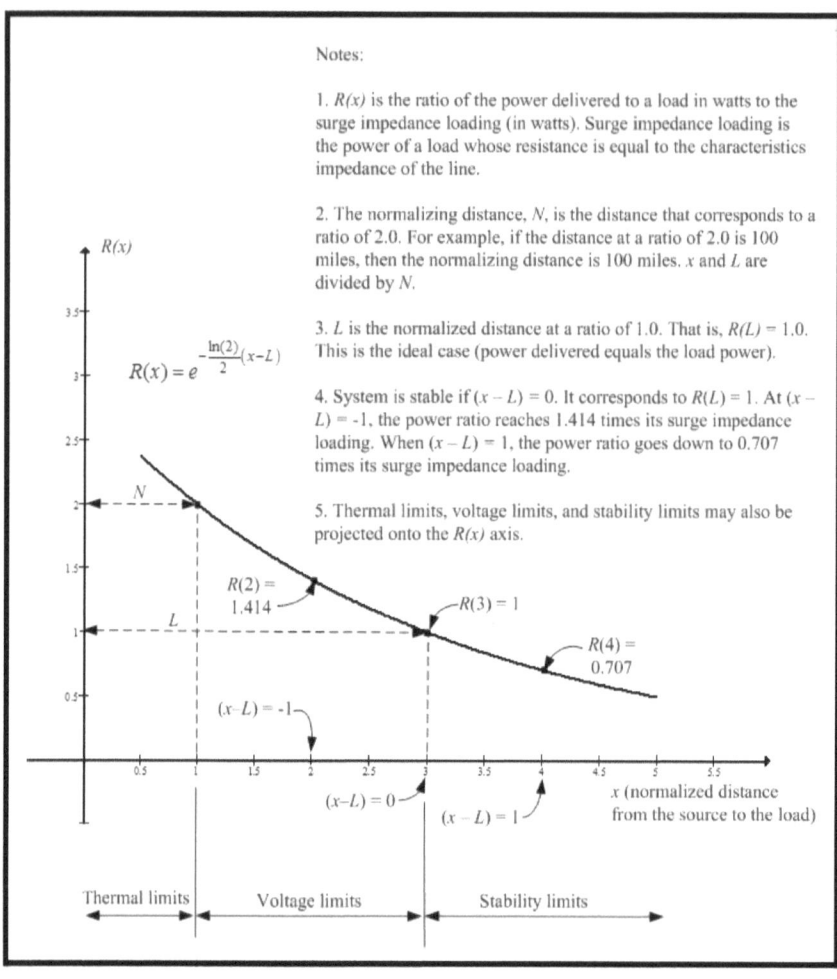

Notes:

1. $R(x)$ is the ratio of the power delivered to a load in watts to the surge impedance loading (in watts). Surge impedance loading is the power of a load whose resistance is equal to the characteristics impedance of the line.

2. The normalizing distance, N, is the distance that corresponds to a ratio of 2.0. For example, if the distance at a ratio of 2.0 is 100 miles, then the normalizing distance is 100 miles. x and L are divided by N.

3. L is the normalized distance at a ratio of 1.0. That is, $R(L) = 1.0$. This is the ideal case (power delivered equals the load power).

4. System is stable if $(x - L) = 0$. It corresponds to $R(L) = 1$. At $(x - L) = -1$, the power ratio reaches 1.414 times its surge impedance loading. When $(x - L) = 1$, the power ratio goes down to 0.707 times its surge impedance loading.

5. Thermal limits, voltage limits, and stability limits may also be projected onto the $R(x)$ axis.

$$R(x) = e^{-\frac{\ln(2)}{2}(x-L)}$$

Figure 3.4 Ratio of actual Loading and Surge Impedance Loading as a Function of Distance

3.2.3   Surge impedance loading in low-voltage (600-volt class) power distribution system

It will be instructive to find out how the concept of surge impedance loading applies to low voltage power distribution system in facilities. For this purpose, the formulas for inductance and capacitance of single-phase line are required. *

The inductance of 2-wire single phase power system with spacing, D, and identical radii, r, is

$$L_{1\phi} = 4 \times 10^{-7} \times \ln\frac{D}{r} \quad \text{(Henry/meter)}$$

Its capacitance is

$$C_{1\phi} = \frac{\pi\varepsilon_0}{\ln\dfrac{D}{r}} \quad \text{(Farad/meter)}$$

Table 3.1 shows some of the characteristics of wires commonly used in low-voltage power distribution system. In steady state, *SIL* is not critical in such applications because their physical lengths are much smaller than the wavelength of the 60 Hz signal. Recall that characteristics impedance is meaningful only on applications where the physical length of a line is comparable to the wavelength of the signal. However, during initial switching of a load, the 60 Hz sinusoidal voltage decays exponentially. A frequency that is comparable to the time constant of the decay occurs in the line and the load. As a result, the output voltage across the load may decrease significantly taking the line across its region of thermal limitation. A circuit breaker protecting the line may trip even if there is no physical fault in any part of the system. Heavy loads, such a 2-horsepower or more motors, are usually equipped with magnetic starters to avoid such tripping. The starters utilize time delay to neutralize the time constant (of the exponential decay).

Table 3.1 Results of the Calculations for the Surge Impedance of a Single Phase Power Cable

Parameter	#8 AWG Conductors	#6 AWG Conductors	#2 AWG Conductors
Geometric mean Distance (meter)	0.00371	0.00467	0.00742
Geometric mean radius (meter)	0.001859	0.002335	0.00371
Inductance (uH/meter)	1.250	1.211	1.119
Capacitance (nF/meter)	8.840	9.176	9.935
Calculated characteristic impedance (ohm)	376.0	363.0	335.6
Surge impedance loading (watt)	115.0	119.0	129.0

Notes:

1. Geometric mean radius taken from Table 5 and Table 8 of the National Electrical Code 2005. Geometric mean distance assumed to be the diameter of the wire.

CHAPTER 4

Instability In Redundant Power Distribution System

Since the advent of computers, more and more customers of the power industry are demanding reliable power system to the point that such a system should be failure proof. Such a failure proof system may be called a fully redundant system. This chapter shows that such a system can be unstable. Alternatives, however, are shown in lieu of such a fully redundant system. While not fully redundant, the alternatives are more energy efficient, less hazardous, and controllable.

4.1 Inherent Instability of a Fully Redundant System –Analysis using Power Flow

Consider a load being fed from two different power sources as shown on Figure 4.1. The power system has full redundancy in the sense that the sources continuously deliver power to the load. If the primary source is lost, the alternate source provides the full power to the load. It will be shown that such a schema can be unstable. Let

S_1 = complex power delivered by first source,

S_2 = complex power delivered by the second source, and

S_3 = complex power consumed by the load.

The following trigonometric identities will be used in the development of the proof:

1. $\cos(x+y) = \cos(x)\cos(y) - \sin(x)\sin(y)$,

2. $\sin(x+y) = \sin(x)\cos(y) + \cos(x)\sin(y)$, and

3. $a\sin(x) + b\cos(x) = \sqrt{a^2 + b^2}\,\sin(x+\varphi)$ where

$\varphi = \tan^{-1}(b/a)$

A complex power may be represented by

$$S = P + jQ$$

where,

P = real part of S, and

Q = imaginary part of S.

When power, with amplitude A, is applied to a load, it experiences a time delay, τ, before reaching the load. The delay is due to the circuit parameters between the source and the load. The real and imaginary parts of the power can therefore be represented as

$$P = A\cos\left(w(t-\tau)\right),\text{ and}$$

$$Q = A\sin\left(w(t-\tau)\right)$$

NOTES:

- THE POWER RATING OF EACH LINE, BUS, AND OTHER COMPONENTS MUST BE GREATER THAN OR EQUAL TO THE SUM OF ALL LOADS AND LOSSES.

- STATEMENT OF FULL REDUNDANCY: WHEN THE PRIMARY SOURCE LOSES POWER, THE ALTERNATE SOURCE MUST PROVIDE POWER TO ALL LOADS WITHOUT INTERRUPTION. CONVERSELY, WHEN THE ALTERNATE SOURCE LOSES POWER, THE PRIMARY SOURCE MUST BE ABLE TO PROVIDE ALL THE POWER.

- THE DESIGN BELOW, HOWEVER, IS UNSTABLE.

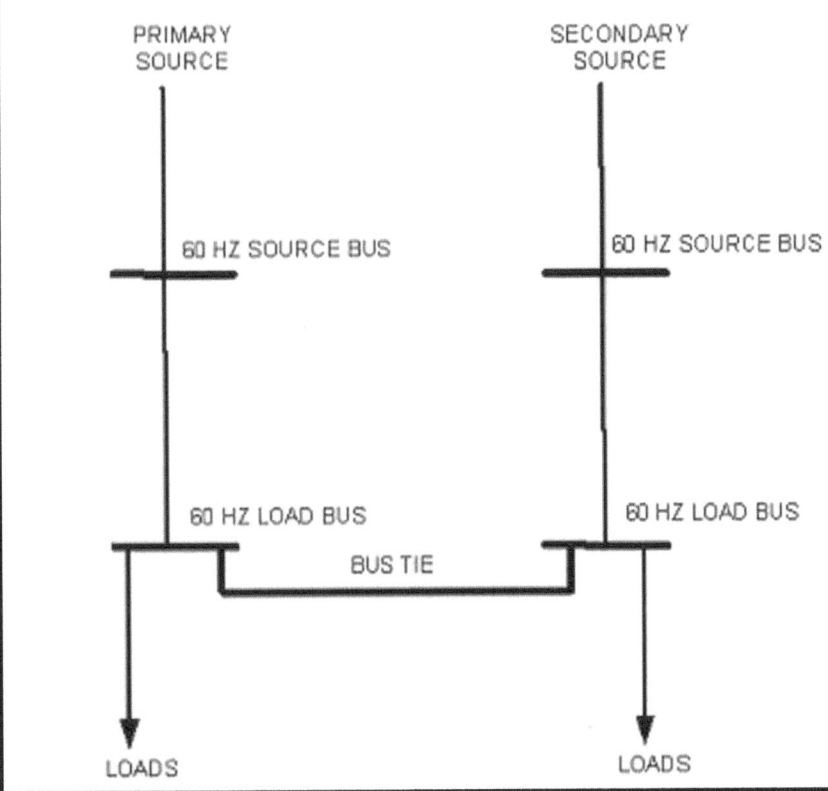

Figure 4.1 Single Line Diagram of a Fully Redundant System

In the following developments, P will be represented by the $re(\ )$ operator, and its Q will be represented by the $im(\ )$ operator. This representation will prevent using too many subscripts to represent power at the source and power at the load. Additionally, P and Q are assumed to have constant amplitude although in reality they decay exponentially.

The real parts of the two sources, using the first identity, may be represented by:

$$re(S_1) = A_1 \cos(w\tau_1)\cos(wt) + A_1 \sin(w\tau_1)\sin(wt)$$, and

$$re(S_2) = A_2 \cos(w\tau_2)\cos(wt) + A_2 \sin(w\tau_2)\sin(wt)$$

Using the second identity, the imaginary components may be represented by:

$$im(S_1) = A_1 \cos(w\tau_1)\sin(wt) - A_1 \sin(w\tau_1)\cos(wt)$$, and

$$im(S_2) = A_2 \cos(w\tau_2)\sin(wt) - A_2 \sin(w\tau_2)\cos(wt)$$

For the load, its real and imaginary components are:

$$re(S_3) = A_3 \cos(w\tau_3)\cos(wt) + A_3 \sin(w\tau_3)\sin(wt)$$, and

$$im(S_3) = A_3 \cos(w\tau_3)\sin(wt) - A_3 \sin(w\tau_3)\cos(wt)$$

The approach in discovering the relationship between a single load and two power sources is to equate the load power to the sum of the loads from two sources. First, simplifying coefficients with two or more factors will help writing the equations and interpreting them later on. Let

$$C = A_3 \cos(w\tau_3)$$, and

$$D = A_3 \sin(w\tau_3)$$

The simpler representation of the load follows:

$$re(S_3) = C\cos(wt) + D\sin(wt)$$ and

$$im(S_3) = C\sin(wt) - D\cos(wt)$$

The sum of the real parts of the two sources is

$$re(S_1) + re(S_2) = A_1\cos(w\tau_1)\cos(wt) + A_1\sin(w\tau_1)\sin(wt)$$
$$+ A_2\cos(w\tau_2)\cos(wt) + A_2\sin(w\tau_2)\sin(wt)$$

or,

$$re(S_1) + re(S_2) =$$
$$\left[A_1\cos(w\tau_1) + A_2\cos(w\tau_2)\right]\cos(wt) + \left[A_1\sin(w\tau_1) + A_2\sin(w\tau_2)\right]\sin(wt)$$

To simplify their representation, let

$$A = A_1\cos(w\tau_1) + A_2\cos(w\tau_2)$$ and

$$B = A_1\sin(w\tau_1) + A_2\sin(w\tau_2)$$

Now,

$$re(S_1) + re(S_2) = A\cos(wt) + B\sin(wt)$$

Similarly,

$$im(S_1) + im(S_2) =$$
$$\left[A_1\cos(w\tau_1) + A_2\cos(w\tau_2)\right]\sin(wt) - \left[A_1\sin(w\tau_1) + A_2\sin(w\tau_2)\right]\cos(wt)$$

or

$$im(S_1) + im(S_2) = A\sin(wt) - B\cos(wt)$$

Under ideal conditions, the real and imaginary parts of the sources and the load are equal. For the real part,

$$A\cos(wt) + B\sin(wt) = C\cos(wt) + D\sin(wt)$$

Doing the same for the imaginary parts,

$$A\sin(wt) - B\cos(wt) = C\sin(wt) - D\cos(wt)$$

Each side of the two equations above can be simplified further by using the third trigonometric identity above. For the real components,

$$\sqrt{A^2 + B^2}\,\sin(wt + \tan^{-1}(A/B)) = \sqrt{C^2 + D^2}\,\sin(wt + \tan^{-1}(C/D))$$

and for the imaginary component,

$$\sqrt{A^2 + B^2}\,\sin(wt + \tan^{-1}(-B/A)) = \sqrt{C^2 + D^2}\,\sin(wt + \tan^{-1}(-D/C))$$

Note that the sum of the two phase shifts in the source, $\tan^{-1}(A/B) + \tan^{-1}(-B/A)$ must equal $\pi/2$. The same is true for the two phase shifts in the load.

The above observation shows that the real and imaginary components must always be 90 degrees or $\pi/2$ radians apart. Hence, an analysis of the real components must yield the same relationships for the imaginary component.

For stability, the source power must be greater than the load power. This means that

$$A^2 + B^2 > C^2 + D^2$$

To find the ideal relationship between the angles corresponding to the phase shifts, first assume a relation. Next, check whether the relation holds. Finally, if the relation is false then its opposite must be true.

Assume

$$\tan^{-1}\left(\frac{A}{B}\right) \le \tan^{-1}\left(\frac{C}{D}\right)$$

or,

$$\frac{A}{B} \le \frac{C}{D}$$

To provide an interpretation of the two inequalities above, let

$$B = D$$

To satisfy the first inequality ($A^2 + B^2 > C^2 + D^2$), A must be greater than C (when $B = D$), or

$$A > C$$

Dividing the last inequality by $B = D$ gives

$$\frac{A}{B} > \frac{C}{D}$$

However, this inequality is the opposite of the original inequality $\frac{A}{B} \le \frac{C}{D}$, and hence a contradiction. That is, A could not be greater than C and less than C at the same time. Therefore, the correct relation must be:

$$\frac{A}{B} > \frac{C}{D}$$

The above development implies that the time constant and amplitude of the load are constrained by the time constants and amplitudes of the sources.

To highlight the above point, examine the coefficients of the cosine term for the sum of the real components of the sources with the coefficients of the cosine term for the real component of the load. Equating them gives,

$$A_1 \cos(wt_1) + A_2 \cos(wt_2) = A_3 \cos(wt_3)$$

Since all the coefficients, frequency, and time constants are all constant, then both sides of the equation must be constant. Furthermore, the time constants of the sources, t_1 and t_2 remains time invariant since they are transmission line constants. The problem, however, will occur when the time constant, t_3 changes when a load is suddenly turned on. Since the transmission line time constants cannot vary, only the coefficients A_1 and A_2 can vary. The above equation, therefore, must be modified to include the difference in the change of amplitudes, $|\Delta A| = |A_1 - A_2|$. That is,

$$A_1 \cos(wt_1) + A_2 \cos(wt_2) + |\Delta A| = A_3 \cos(wt_3) \text{ (during transient)}.$$

If t_3 is very small (increasing its cosine term) then $|\Delta A|$ must be large.

4.2   Other Methods of Analysis

The previous paragraph used complex power, some trigonometric identities, and time constants of the transmission lines and the load to arrive at the conclusion that a fully redundant system as shown on Figure 4.1 could be unstable. Three other methods of analysis are shown below to verify such instability.

4.2.1 Using the conventional circuit analysis

A circuit analysis approach of the problem may be made by considering two independent voltage sources in parallel with a single load. Use superposition theorem to reduce the circuit complexity. Next, obtain the transfer function assuming the same input voltage for the two sources. The result will be a transfer function whose number of zeros and poles is much more than the transfer function with one source.

4.2.2 Using propagation constants

Yet, another approach to obtain a feel of the complexity of the problem is to use the solution of the partial differential equation of a transmission line. When there are two transmission lines, there are two propagation constants. Each constant has real and imaginary parts. Furthermore, the solution has forward and backward traveling waves that depend on two initial conditions of voltage. In all, the number of parameters is at sixteen. Compare this with a single transmission line whose complexity consists of four parameters – real and imaginary parts of the propagation constant and two initial conditions of voltage. Doubling the number of transmission lines increased its complexity four times.

4.2.3 Approach using the convolution theorem

The convolution theorem relates the output of a system with its input and transfer function. In the frequency domain (with Laplace variable s), the output is the product of the input and the transfer function. Conversely, the transfer function is the ratio of the output and the input.

In the redundant system of Figure 4.1, there are two sinusoidal inputs. The output is exponentially decaying sinusoidal wave. Exponential decay is due to the attenuation constant of the transmission line.

As chapter 3 shows, when a wave travels along a transmission line, it experiences exponential decay due to the attenuation constant (real component) of the propagation constant. From this information, the transfer function of the system shown on Figure 4.1 is

$$H(s) = \frac{k_1 (s - \alpha_1)(s^2 + \omega^2)}{s\left((s - \alpha_1)^2 + \omega^2\right)} + \frac{k_2 (s - \alpha_2)(s^2 + \omega^2)}{s\left((s - \alpha_2)^2 + \omega^2\right)}$$

Expanding the factors in the denominator shows that some of its coefficients are negative. It implies that the transfer function is unstable as shown in [22]. Hence, the system represented on Figure 4.1 is unstable as well.

4.3   Improving the Response

The instability of the fully redundant system maybe reduced by re-routing one of the redundant cable, installing a synchronizing relay, or implementing the redundancy at the DC level (for those applications whose outputs are DC voltage such as in computer systems). These options are shown on Figures 4.2, 4.3, and 4.4.

4.3.1 Re-routing a cable and installing a circuit breaker

Re-routing a redundant cable, as shown on Figure 4.2, forces the load to see one transmission line only. Hence, it will see one propagation constant only. The circuit breaker provides the bus tie between the two sources. It is normally open and closes only when the primary power is lost.

4.3.2 Using a synchronizing relay at the load level

Figure 4.3 uses a synchronizing relay to connect exactly one source to the load if one of the two sources is disabled. The relay may be used to close relay contacts in automatic transfer switch or a circuit breaker.

NOTES:

- THE POWER RATING OF EACH LINE BUS, AND OTHER COMPONENTS MUST BE GREATER THAN OR EQUAL TO THE SUM OF ALL LOADS AND LOSSES.

- WHEN THE PRIMARY SOURCE LOSES POWER THE CIRCUIT BREAKER CLOSES AND ACTIVATES THE SECONDARY SOURCE A SHUNT TRIP DEVICE MUST BE USED WHEN CLOSING THE BREAKER

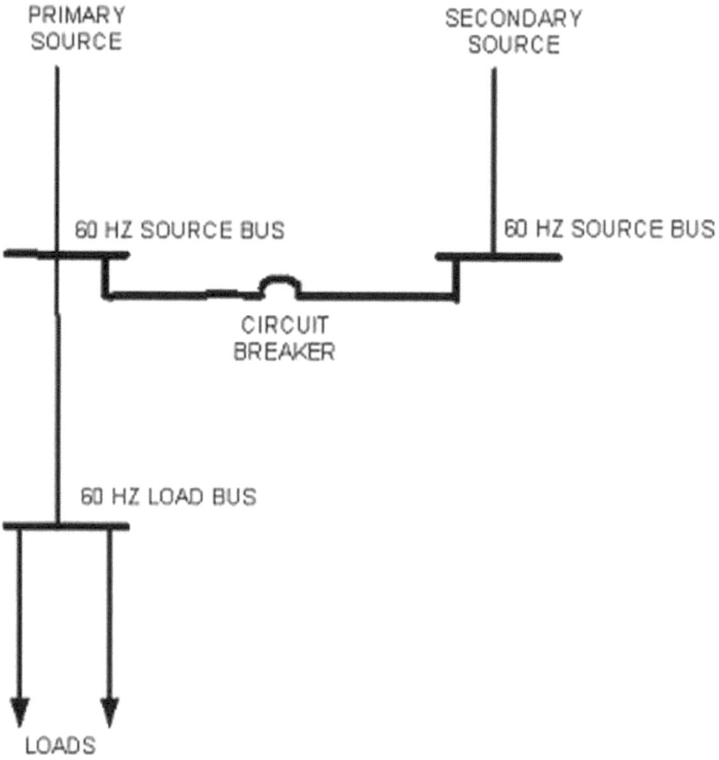

PRIMARY
SOURCE

SECONDARY
SOURCE

60 HZ SOURCE BUS

60 HZ SOURCE BUS

CIRCUIT
BREAKER

60 HZ LOAD BUS

LOADS

**Figure 4.2 Single Line Diagram of a System with a Bus Tie
at the Source**

NOTES:

- THE POWER RATING OF EACH LINE BUS, AND OTHER COMPONENTS MUST BE GREATER THAN OR EQUAL TO THE SUM OF ALL LOADS AND LOSSES.

- WHILE INSTALLING A SYNCHRONIZING RELAY MAY NOT PROVIDE FULL REDUNDANCY, IT MAKES THE SYSTEM STABLE BY FORCING THE LOAD TO SEE ONE POWER ONLY AND ONE PROPAGATION CONSTANT

- THE ADVANCE(CLOSING) ANGLE OF THE SYNCHRONIZING RELAY IS

$$\theta = 360st$$

WHERE

θ = closing angle in degrees,

s = slip frequency in cycles per second, and

t = the closing of the breaker in second.

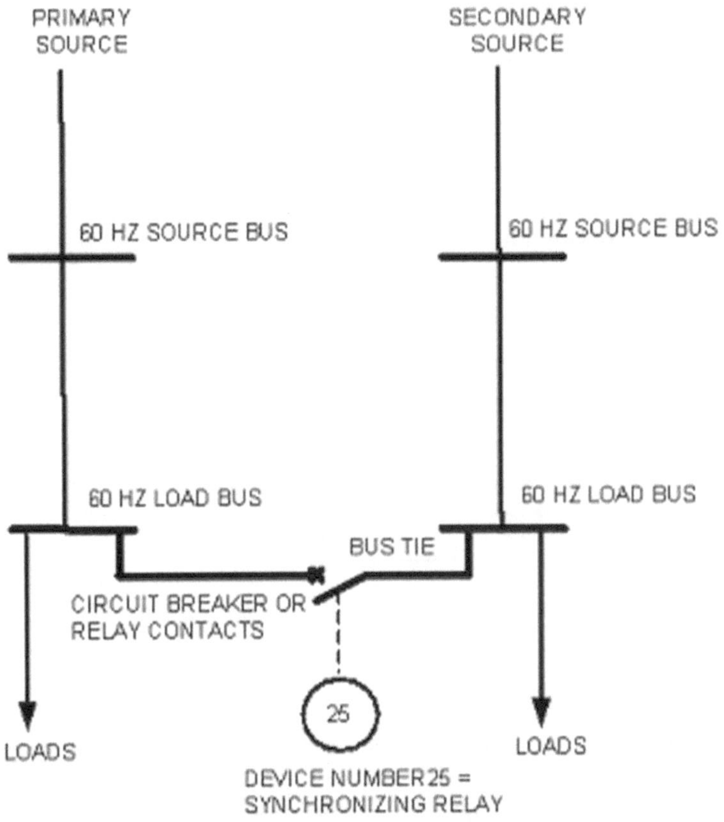

PRIMARY
SOURCE

SECONDARY
SOURCE

60 HZ SOURCE BUS

60 HZ SOURCE BUS

60 HZ LOAD BUS

60 HZ LOAD BUS

BUS TIE

CIRCUIT BREAKER OR
RELAY CONTACTS

25

LOADS

LOADS

DEVICE NUMBER 25 =
SYNCHRONIZING RELAY

Figure 4.3 A System with Synchronizing Relay

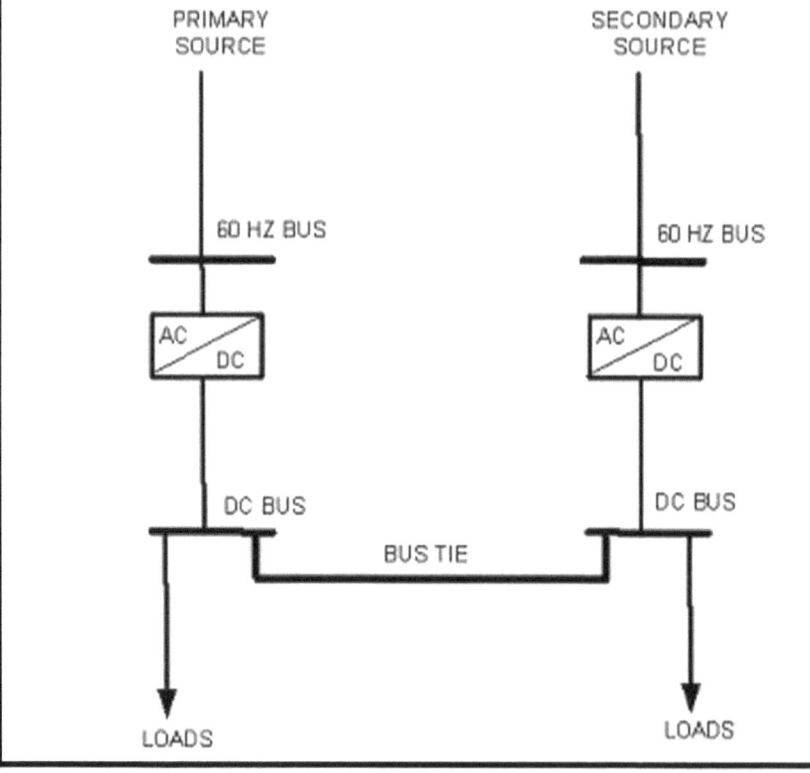

NOTES:

- THE POWER RATING OF EACH LINE BUS, AND OTHER COMPONENTS MUST BE GREATER THAN OR EQUAL TO THE SUM OF ALL LOADS AND LOSSES.

- THE DC BUS MUST HAVE ITS OWN PROTECTION IN CASE OF A FAULT AT THE LOAD SIDE IT IS POSSIBLE THAT 60 HZ PROTECTION MAY NOT TRIP WHEN SUCH A FAULT OCCURS

- IF THE LOADS ARE ALL DC, IT PAYS TO INSTALL THE REDUNDANCY AT THE DC LEVEL AND NOT AT THE 60 HZ LEVEL

PRIMARY SOURCE

SECONDARY SOURCE

60 HZ BUS

60 HZ BUS

AC DC

AC DC

DC BUS

DC BUS

BUS TIE

LOADS

LOADS

Figure 4.4 A System with Full Redundancy at the DC Output

4.3.3   Implementing redundancy at the DC level

The last option requires DC loads. This is shown on Figure 4.4. DC power is obtained from AC power by rectification. The transfer function of the system may be represented by

$$H(s) = \frac{s^2 + \omega^2}{s^2 \left(as^2 + bs + 1\right)}$$

Since all coefficients of the denominator are all positive then $H(s)$ is stable.

4.4   Remark on the Unifying Role of Mathematics and the Power of the Laplace Transform

The above sections show the important role of mathematics, specifically the Laplace transform, in understanding the stability of a system. Not only is Laplace transform useful in the analysis but in design as well.

Consider a system consisting of input as a function of time, a hardware consisting of electrical elements such as resistor, inductor, and capacitor, and an output, which is also a function of time. Oftentimes, the task of an engineer is to design the hardware given the input and the output.

The convolution theorem may serve as the starting point of such a design.

As an example, consider designing a system that can sense a fault and open a circuit breaker when such a fault occurs. Visualize the input as exponentially decaying voltage when a fault occurs. The input voltage and its corresponding Laplace transform is

$$e^{-at}u(t) \longleftrightarrow \frac{1}{s+a}$$

The output function is a unit step of the form:

$$u(t-\tau) \longleftrightarrow \frac{e^{-\tau s}}{s}$$

Simplifying the exponential function

$$e^{-\tau s} = \frac{1}{e^{\tau s}}$$

and using the approximation

$$e^x = 1 + x + \frac{x^2}{2} + \dots$$

the ratio of the output and input, or transfer function of the hardware, is

$$H(s) = \frac{k(s+a)}{s(s^2 + qs + p)}$$

Once $H(s)$ is known, it can be realized or implemented. References [16] to [22] may be used for such a purpose.

CHAPTER 5

Testing of Electronic Systems

An electronic system is usually tested to (1) verify its performance; or, (2) find a fault or cause of its failure. In both cases, signals are applied at the inputs of a system and the output monitored. For the second case, however, searching for the true cause of a fault requires probing intermediate points in a subsystem. This chapter shows the technique of tracing a fault from the output back to the input.

5.1   General Technique of Tracing a Fault

Figure 5.1 shows a system consisting of two subsystems. Each subsystem consists of DC power supply, input lines, output lines, and grounding system. A subsystem can be an analog, digital, RF, microwave or other circuits.

Each input line or output line in Figure 5.1 is actually composed of three lines as shown on Figure 5.2. They are the signal line, the return line, and the ground line. For simplicity, a line can be a wire, a copper trace on a printed circuit board assembly, or any interconnecting medium.

Figure 5.3 is an illustration of a simple subsystem. It consists of an operational amplifier and set-reset (SR) flip flop. The figure illustrates the signal source, the three different lines, power supplies, and grounding points of each device. Note that all grounding points of each circuit are connected to a single common ground point (marked "A"). Not shown on the figure are the connections of this common point with the chassis and the facility grounding subsystem.

The general technique of tracing a fault is shown on Figures 5.4 through 5.8. When a fault is observed at the output, the possible cause of the fault can be any of the parts external to the output. A boundary line encloses these parts.

To trace the cause of the fault back, the input lines of the last device are probed and its signals compared to known good signals. Consider that the top input line (indicated by sub-fault 1) is also faulty. Then, all the parts that can possibly affect this line are enclosed by another boundary line indicating the possible cause of the sub-fault. The procedure continues until a ground line is probed. If it is still faulty, then the grounding subsystem (enclosed by another boundary line for sub-fault 4) may be defective. A rigorous testing of the grounding system may be required at this time.

Notice the general pattern of tracing a fault. It starts from a large boundary and ends in a smaller boundary.

5.2   Tracing a Fault within a Boundary

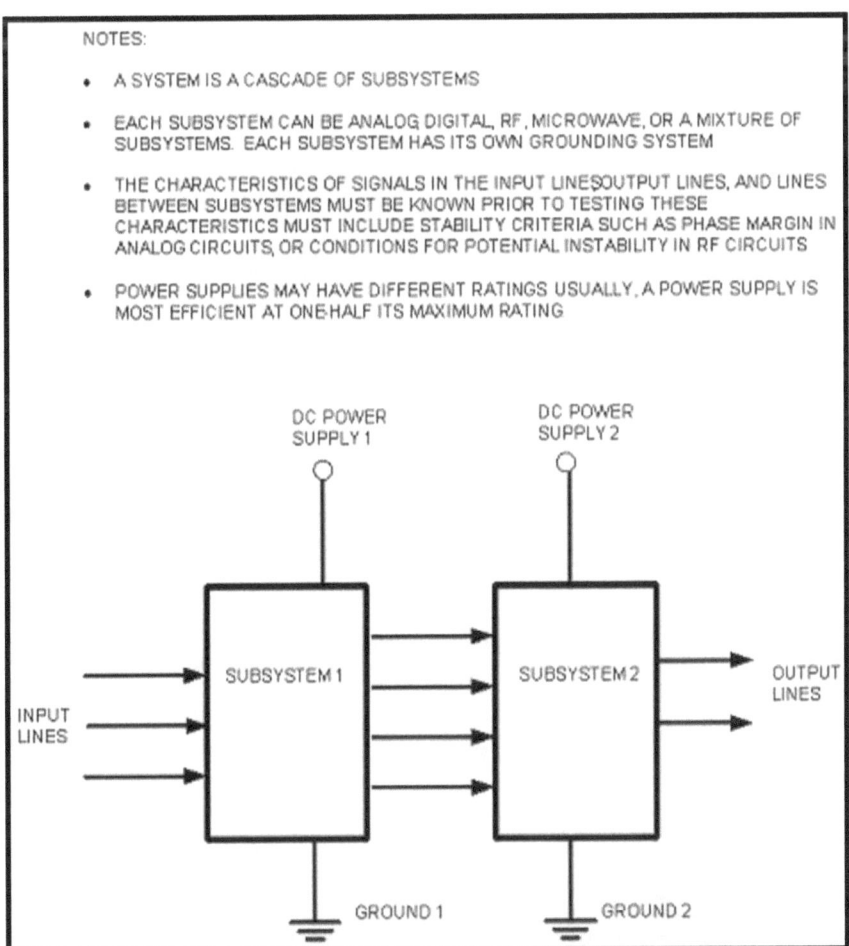

NOTES:

- A SYSTEM IS A CASCADE OF SUBSYSTEMS

- EACH SUBSYSTEM CAN BE ANALOG DIGITAL, RF, MICROWAVE, OR A MIXTURE OF SUBSYSTEMS. EACH SUBSYSTEM HAS ITS OWN GROUNDING SYSTEM

- THE CHARACTERISTICS OF SIGNALS IN THE INPUT LINES OUTPUT LINES, AND LINES BETWEEN SUBSYSTEMS MUST BE KNOWN PRIOR TO TESTING THESE CHARACTERISTICS MUST INCLUDE STABILITY CRITERIA SUCH AS PHASE MARGIN IN ANALOG CIRCUITS, OR CONDITIONS FOR POTENTIAL INSTABILITY IN RF CIRCUITS

- POWER SUPPLIES MAY HAVE DIFFERENT RATINGS USUALLY, A POWER SUPPLY IS MOST EFFICIENT AT ONE-HALF ITS MAXIMUM RATING

DC POWER SUPPLY 1

DC POWER SUPPLY 2

SUBSYSTEM 1

SUBSYSTEM 2

INPUT LINES

OUTPUT LINES

GROUND 1

GROUND 2

Figure 5.1 A System Consisting of Subsystems

NOTES:

- A SIGNAL LINE, RETURN LINE, OR A GROUND LINE MAY BE A WIRE COPPER TRACE AS IN PRINTED CIRCUIT BOARD OR A METAL PLANE AS IN MICROWAVE ASSEMBLIES.

- A SIGNAL LINE OR A RETURN LINE MAY CARRY A PREDETERMINED HIGH OR LOW FREQUENCY SIGNAL. THE LINES MUST HAVE SUFFICIENT BANDWIDTH TO MAINTAIN THE SPECTRAL PURITY OF THE SIGNAL.

- A GROUND LINE MAY CARRY UNPREDICTABLE SIGNALS ALTHOUGH THEY LOOK RANDOM, GROUND SIGNALS HAVE PREDICTABLE CHARACTERISTICS SUCH AS MINIMUM AND MAXIMUM PEAK-TO-PEAK VALUES, ROOT MEAN SQUARE (RMS) VALUES, OR POWER SPECTRUM.

- ANY SIGNAL, RETURN, OR GROUND LINE CAN AFFECT THE PERFORMANCE OF THE SYSTEM.

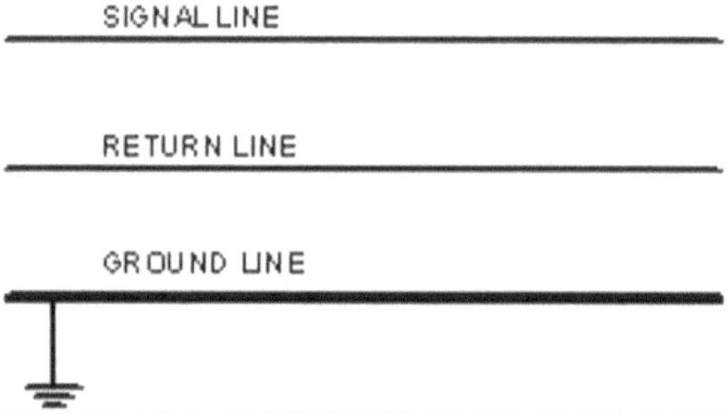

SIGNAL LINE

RETURN LINE

GROUND LINE

Figure 5.2 Input or Output Line Typically Consists of Three Lines

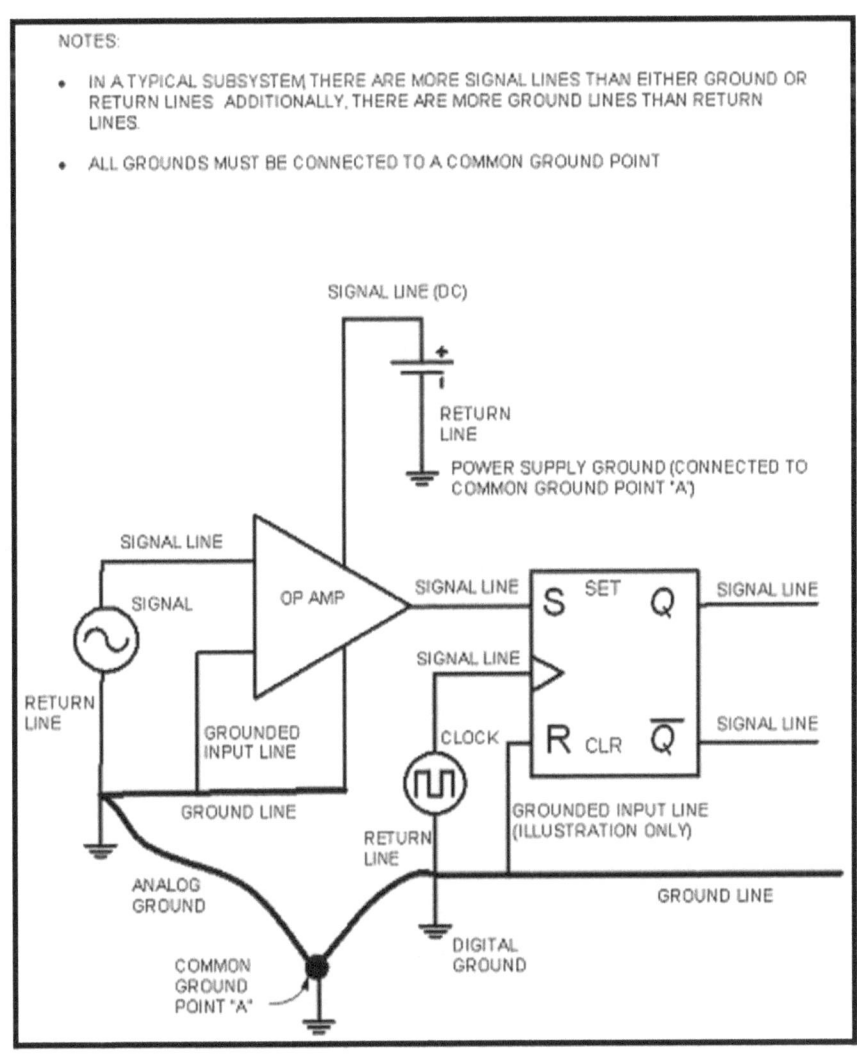

Figure 5.3 A Simple Subsystem Showing the Location of the Signal, Return, and Ground Lines

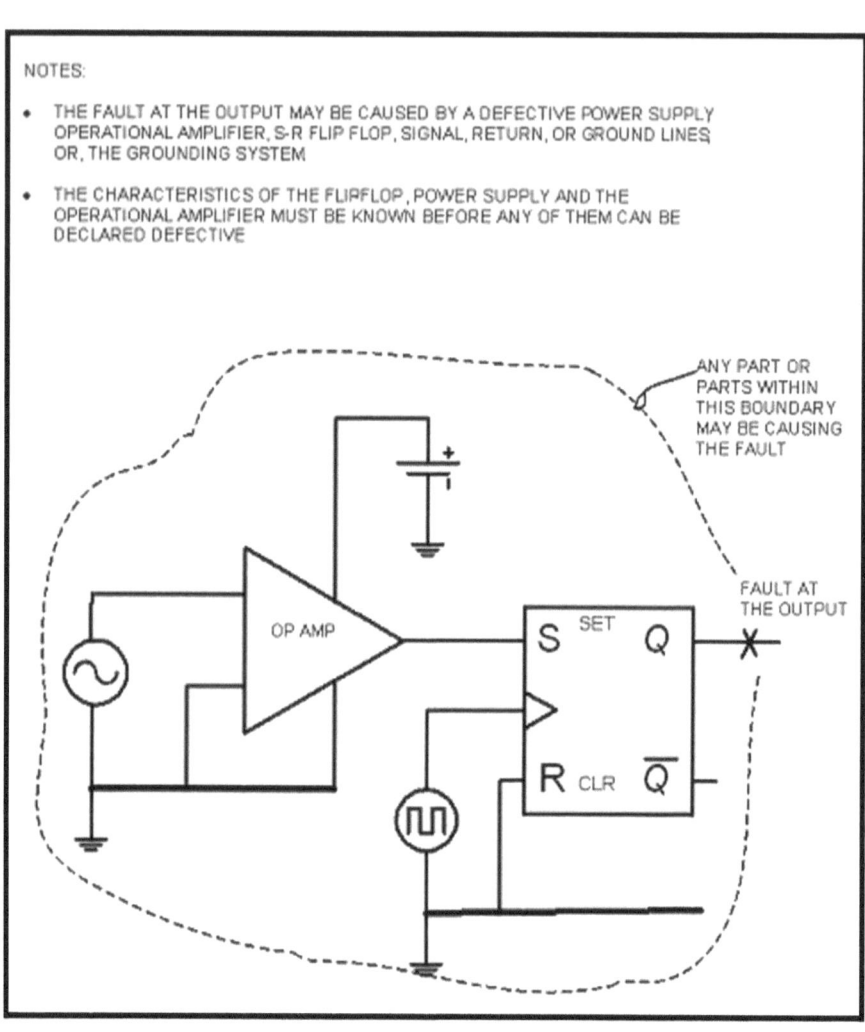

**Figure 5.4 A System with a Fault on the Output and the
Boundary of its Possible Defective Parts**

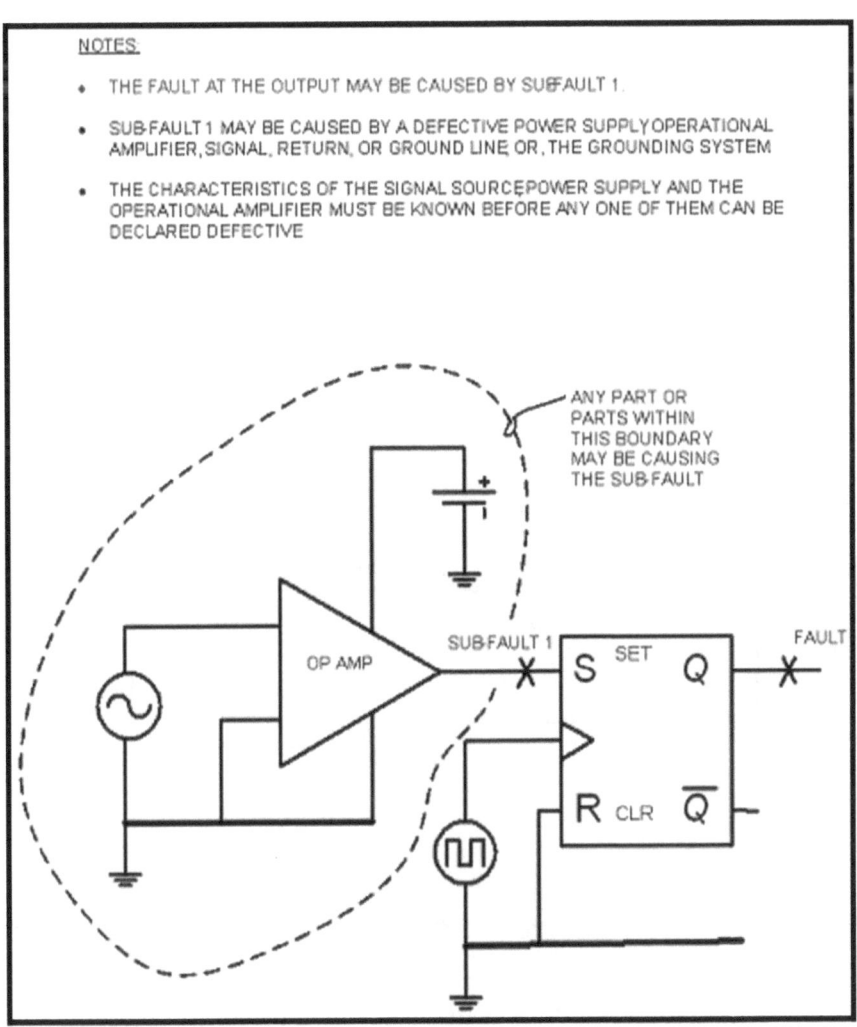

Figure 5.5 Sub-Fault 1 Causing the Fault at the Output

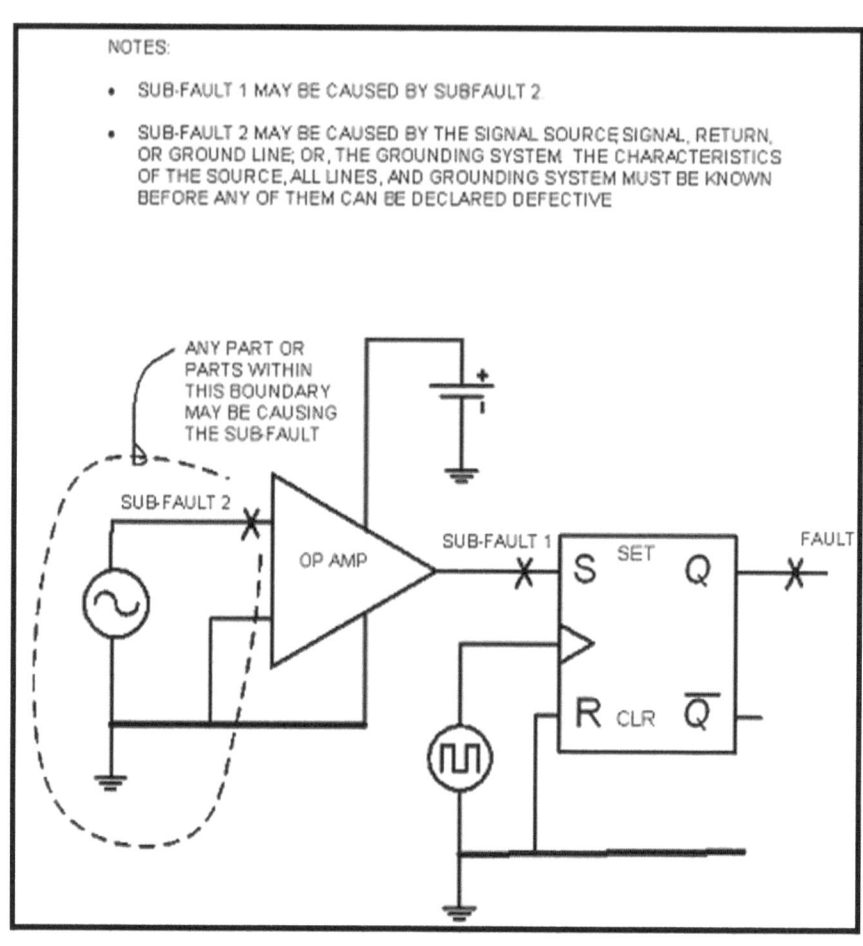

Figure 5.6 Sub-fault 2 Causing Sub-fault 1

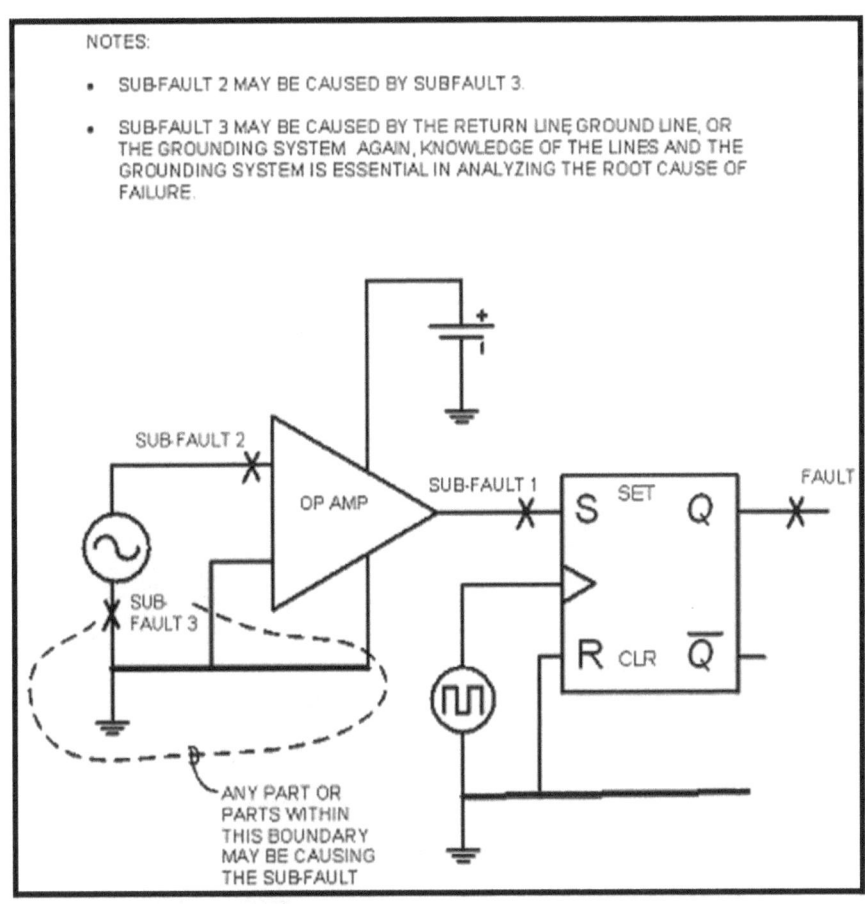

Figure 5.7 Sub-fault 3 Causing Sub-fault 2

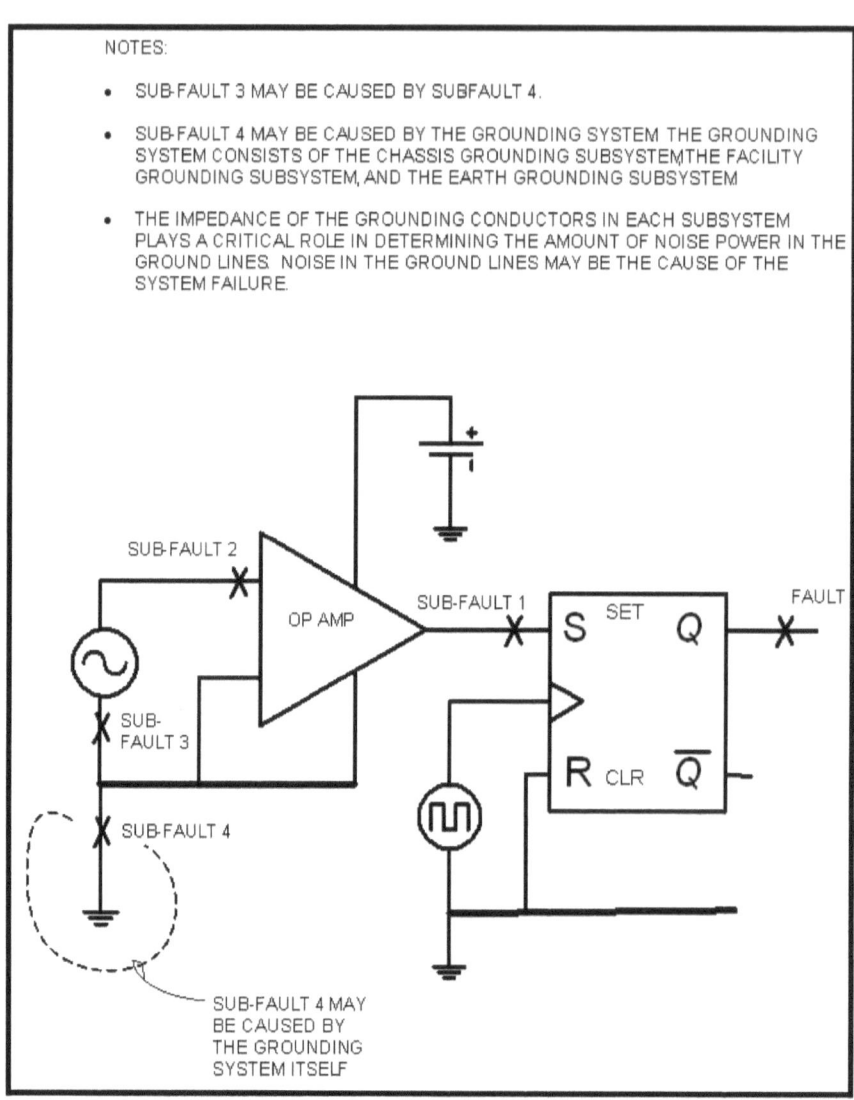

Figure 5.8 The Grounding System Causing Sub-fault 3

The above section shows how a fault is traced from one boundary to another boundary. Within a boundary, a fault may be traced according to the following hierarchy:

1. DC power supply lines,
2. devices,
3. signal lines,
4. return lines, and
5. ground lines.

The above hierarchy is not without rationale. A DC power supply provides DC bias to several circuits. If a fault on the DC power supply occurs then several parts can be eliminated quickly as the possible cause of fault. Furthermore, a DC power supply has a constant waveform. There is no need to check for its bandwidth or other signal characteristics.

A line-to-ground fault is, perhaps, the most common type of fault in a DC power supply. It can be easily checked by probing the node of a circuit connected to a supply.

A device may have a short circuit or open circuit in one of its terminals. Furthermore, for digital devices, there is so-called stuck-at-one (s-a-1) or stuck-at-zero (s-a-0) fault. Poor grounding often causes these faults.

Next in hierarchy are the signal lines. A signal consists of frequency, rise and fall times, phase shift or delay, or attenuation. These characteristics are oftentimes enough to determine if a signal line is good or bad.

Return lines carry the same current as the signal lines. Their voltages, however, are near zero since they are grounded.

When a fault could not be isolated to a DC power supply, device, signal line, or return line, the last choice is usually the grounding system. Poor grounding system account for the majority of faults in electronic systems.

5.3   Knowledge Base of Signals and Circuits

The last section shows the importance of characterizing a signal when tracing a fault. Equally important is the knowledge of circuits. A circuit produces a signal. Defective circuits will produce defective signals. Oftentimes, there is a need to reconstruct the structure of a system when its output consistently gives a bad signal. Chapter 6 shows how to reconstruct a circuit for which the specifications are unknown. Chapter 7 shows how a signal may be characterized.

CHAPTER 6

Reconstructing the Low Pass Filter of a DC Power Supply

This chapter shows how an unknown low pass filter of a DC power supply may be characterized. The basis of the characterization is the convolution theorem.

As stated in Chapter 1, switching in the DC power supplies is the culprit of much of the noise not only in electronic systems but also in the facility power system as well. Low pass filters are used to minimize such noise. Unfortunately, there are filters that are poorly designed. Specifically, if the filter has inadequate order or uses a grounding system other than single point grounding system, the filter may exhibit unacceptable noise floor.

In this chapter, the author reconstructed the frequency response of a low pass filter whose specifications are not accurately known. The goal is to find its -3 dB cutoff point. Additionally, the author simulated the response of the filter in time domain. Variations in the values of inductors or capacitors are shown in the response. Voltage noise sources are also applied across critical parts of the filter to verify if, indeed, they will show in the response.

6.1 Design of a Typical Power Supply

The design of a typical power supply uses a pulse width modulation to maximize signal to noise ratio. In such a schema, the information is in the widths of the pulses. A high power field effect transistor does the actual switching of power from the building power source. Built-in microprocessor precisely controls the widths of the pulses.

Low pass filter attenuates the backward propagating noise from the rectifier to the AC power source. While such filters may be adequate when a power supply is used alone paralleling two or more power supplies may prove that the order of such a filter may be inadequate.

6.2 Reconstructing the Frequency Response of an Unknown Low Pass Filter in a DC Power Supply

Figure 6.1 shows the structure of a power supply that the author reconstructed from its schematic diagram. Note the use of different grounding systems used in the supply.

The best way to reconstruct the frequency response of the filter is to use an oscillator in the form of a portable signal generator that can generate sinusoids at different frequencies. Oscillator output is applied across the sampling network and ground. The response of the filter is measured at the input or the facility power input.

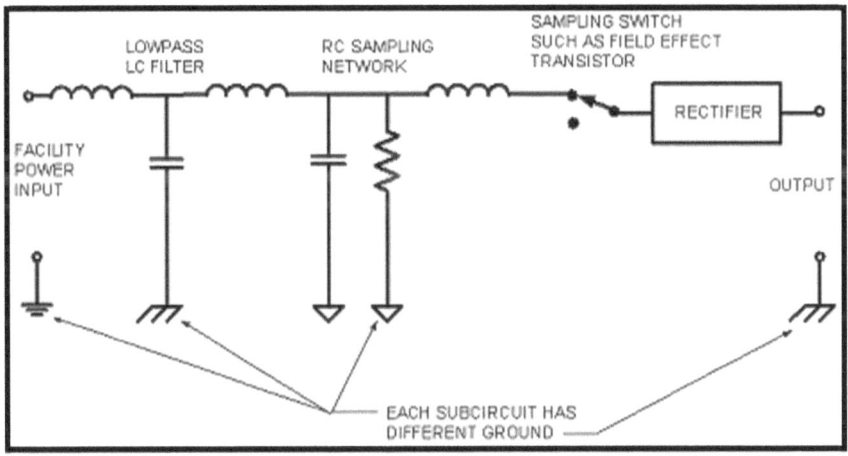

**Figure 6.1 A Typical Power Supply with Different
Grounding Points**

For verifying the frequency response of the filter relative to a grounding system, two grounding schemas are used. They are:

1. using the original multiple grounding system as shown on Figure 6.1, and

2. modifying the original grounding system to form a single point grounding system as shown on Figure 6.2.

Figure 6.3 shows the frequency response of the low pass filter using each grounding schema.

Notice that the -3 dB cutoff frequency of the low pass filter in Figure 6.1 is narrower than that of Figure 6.2. Furthermore, Figure 6.1 has a gain of about -20 dB at the switching frequency of the rectifier, which is about 50 KHz. In contrast, Figure 6.2 has a gain of -40 dB at the same frequency. The noise floor of Figure 6.2 is, therefore, lower.

Since the reactance of an inductor increases with higher inductance, and the reactance of a capacitor decreases with higher capacitance, an increase in value of either element from Figure 6.1 or 6.2 will move the -3 dB cutoff point to the left. These inductors or capacitors are shown on Figure 6.4.

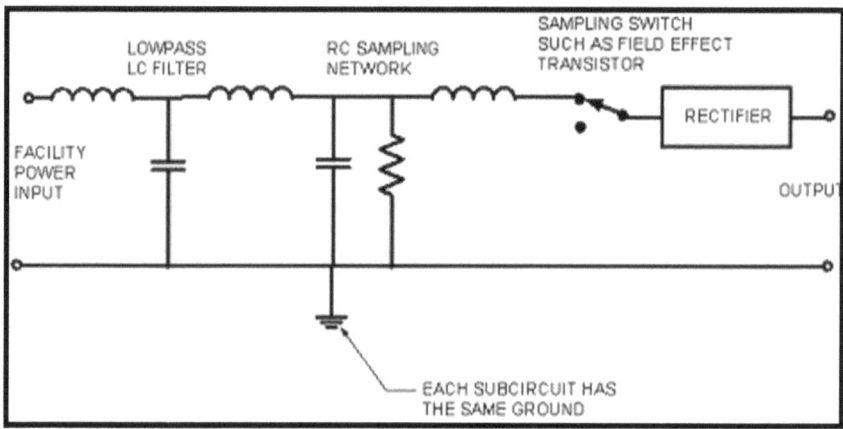

Figure 6.2 The Power Supply with Single Point Grounding

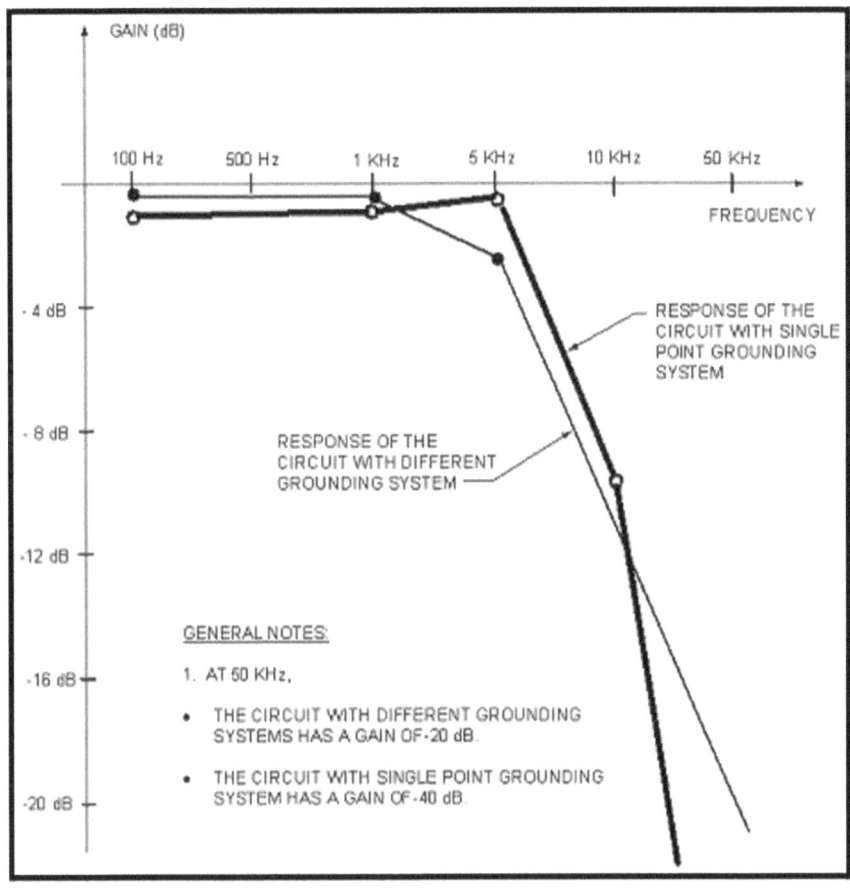

Figure 6.3 Frequency Response of the Filters

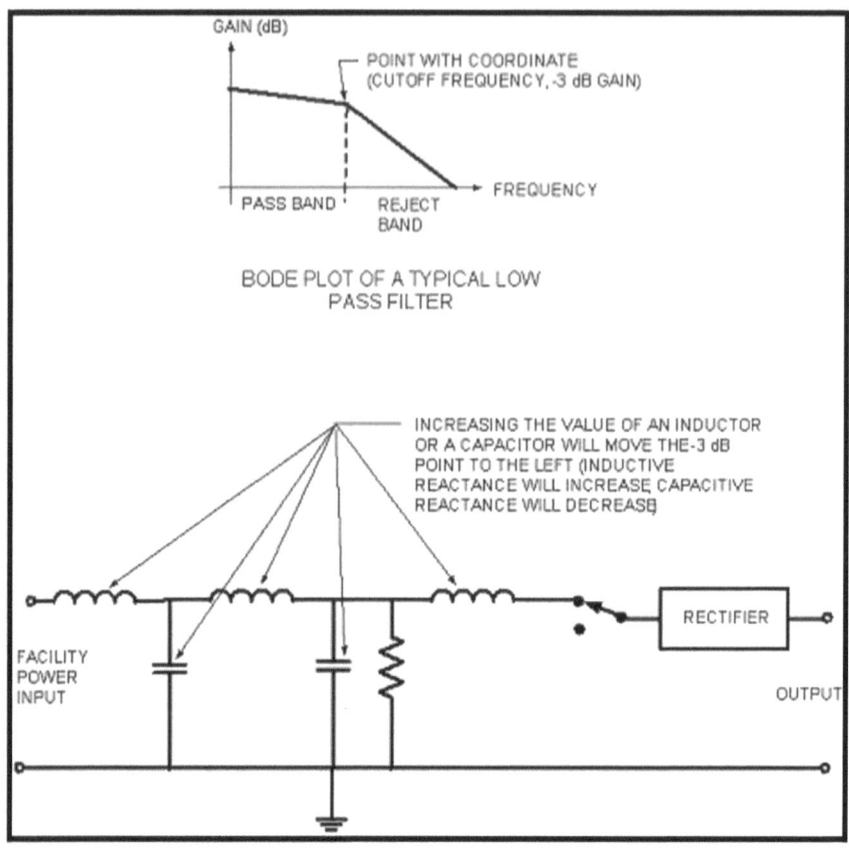

Figure 6.4 Variations of Inductor and Capacitor in the Low Pass Filter of the Power Supply

6.3 Computer Simulations of the Reconstructed Low Pass Filter in Time Domain

In addition to actually measuring the frequency response of the filter, simulation software was used to study its response in time domain. Specifically, the time domain response shows the variation of the output relative to:

1. variation in the size of inductors or capacitors,

2. paralleling two power supplies,
3. injecting noise at the output of the rectifier, and
4. injecting noise between two grounding points.

The noise is in the form of a repetitive pulse train. A repetitive pulse has odd harmonics and is the sum:

$$\frac{4V(t)}{\pi}\left[\sin(2\pi f) + \frac{1}{3}\sin(2\pi(3f)) + \frac{1}{5}\sin(2\pi(5f)) + \frac{1}{7}\sin(2\pi(7f)) + ...\right]$$

where,

$V(t)$ = peak-to-peak voltage of the voltage in volt, and

f = frequency in Hz.

Figures 6.5 through 6.9 are the results of the simulation. Two waveforms are shown on each simulation. The first waveform (lighter and smooth) is the normalized 60 Hz facility power input. The other waveform is the output of the filter (darker and with oscillations). Values of capacitors and inductors determine the attenuation of the filter's output. A time delay of 18 milliseconds was used to trigger the repetitive pulse. From zero to 18 milliseconds the circuit sees the 60 Hz facility power source only. Thereafter, it sees not only the 60 Hz but other harmonics as well.

Some of the conclusions that can be derived from the simulations are:

1. as the number of loads increases the amplitude of the output decreases,
2. injected noise between two grounding points appears as ripples in the output, and
3. decreasing the inductance of the facility power source improves the response.

The first conclusion is a consequence of the power equation $P = VI$. As the number of loads increases so is the current. Increased current causes the voltage drop across the wires feeding the load to increase. Hence, the amplitude of the output must decrease.

The second conclusion is a consequence of the superposition theorem.

For the third conclusion, lower inductive reactance of the facility source moves the -3 dB point to the right creating wider pass band. As mentioned before, the wider pass band lowers the noise floor of the filter at the switching frequency.

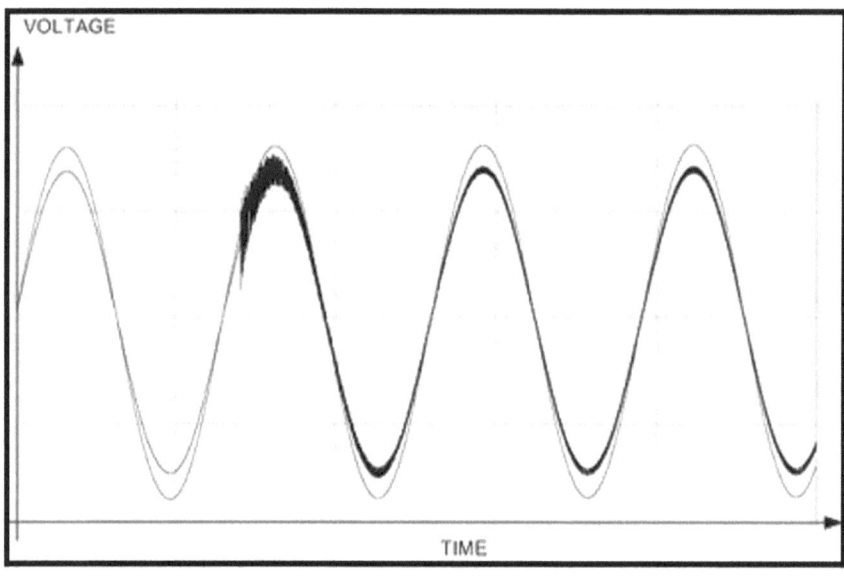

Figure 6.5 Reduced Capacitance and Increased Inductance

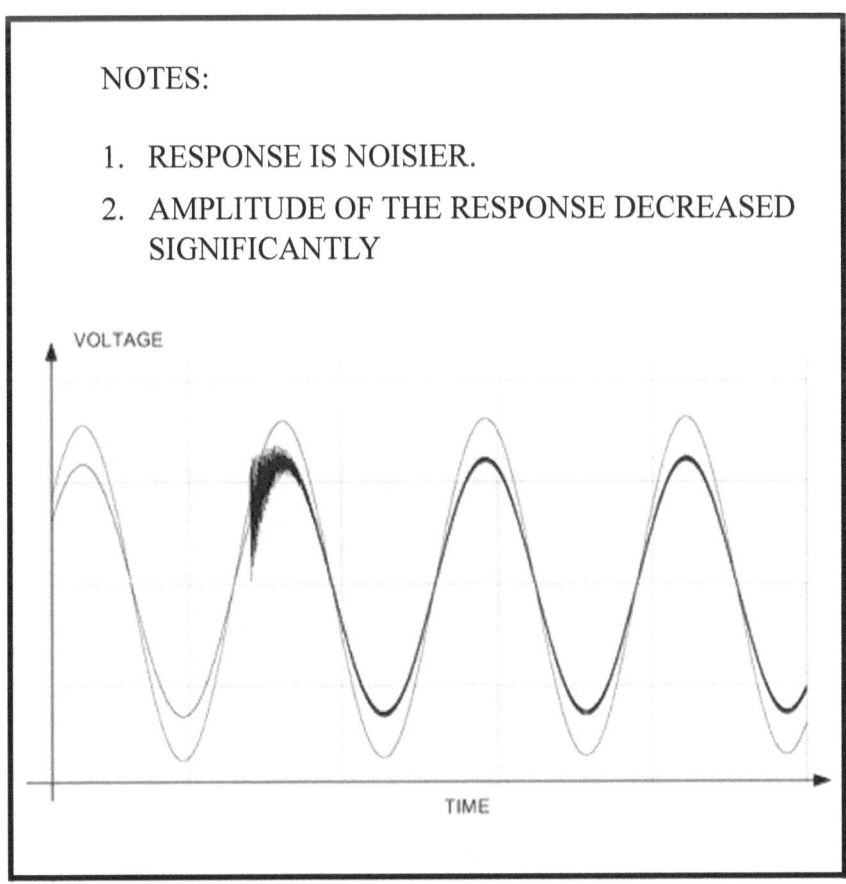

NOTES:

1. RESPONSE IS NOISIER.

2. AMPLITUDE OF THE RESPONSE DECREASED SIGNIFICANTLY

Figure 6.6 Conditions Similar to Figure 6.5 in Addition to Doubling the Number of Power Supplies

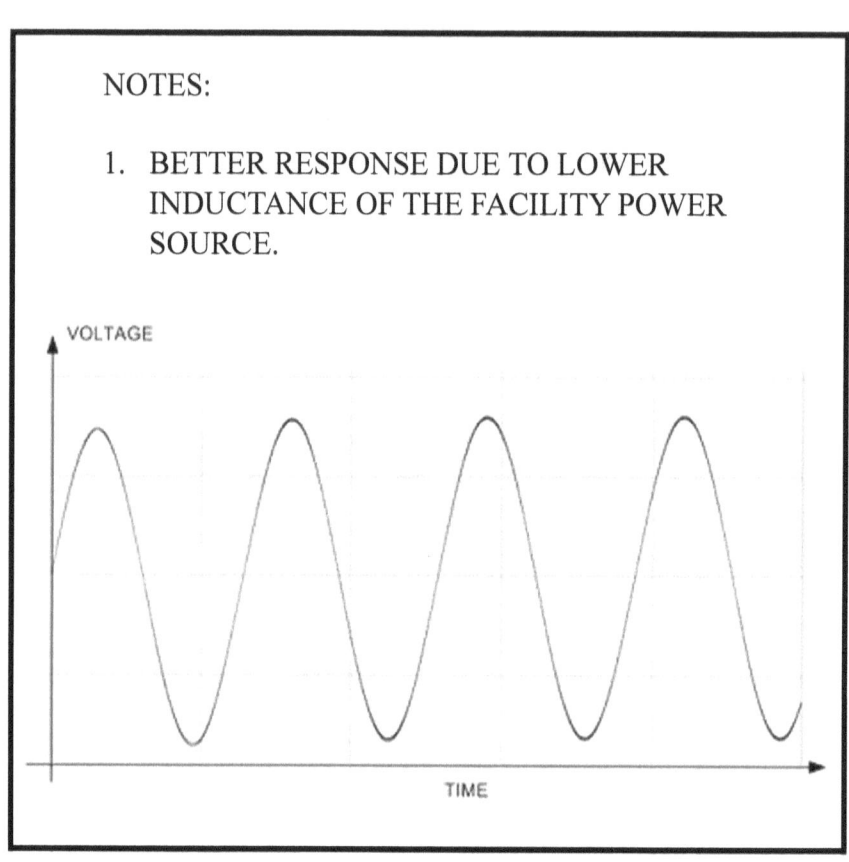

NOTES:

1. BETTER RESPONSE DUE TO LOWER INDUCTANCE OF THE FACILITY POWER SOURCE.

Figure 6.7 Reduced Inductance of the 60 Hz Source

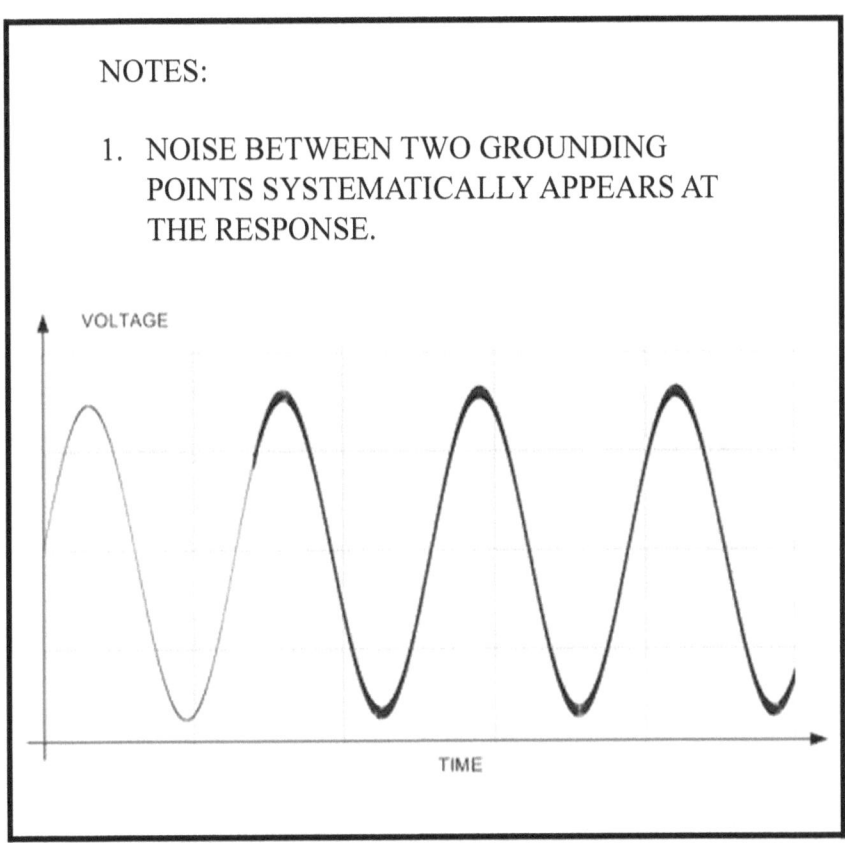

NOTES:

1. NOISE BETWEEN TWO GROUNDING
   POINTS SYSTEMATICALLY APPEARS AT
   THE RESPONSE.

VOLTAGE

TIME

**Figure 6.8 Similar to Previous but with Noise Generated
between two Grounding Points**

NOTES:

1. INCREASED CAPACITANCE MOVES THE -3dB POINT TO THE LEFT DECREASING THE PASS BAND OF THE FILTER.

2. AMPLITUDE OF THE RESPONSE DECREASED SIGNIFICANTLY.

3. WIDTHS OF THE NOISE INCREASED.

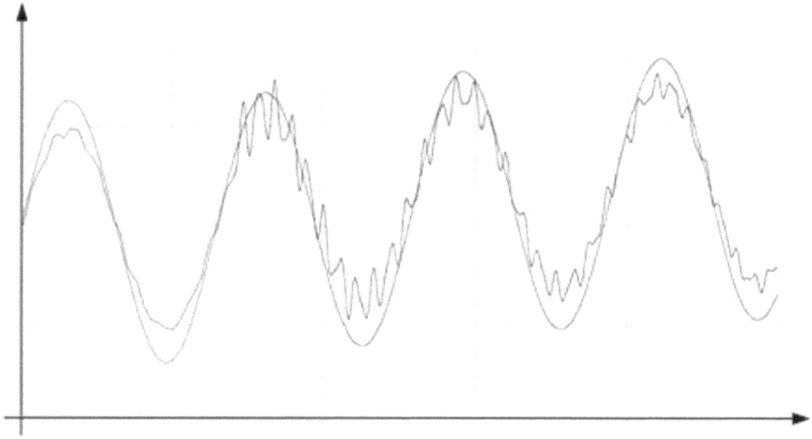

Figure 6.9 Exaggerated the Size of the Capacitors

CHAPTER 7

Experiments on Grounding System and Noise

As described in chapter 5, tracing a fault may require characterization of a signal whether the signal is in the signal line itself, in a return line, or in a ground line. A grounding noise voltage is a signal. To the casual observer, a grounding noise voltage appears to have random frequency, phase, or amplitude. In reality, it is not completely random.

In this chapter, several experiments were performed to study the behavior of grounding noise voltage relative to materials and circuits for grounding. These materials include wires, ferrite beads, capacitors, and flat braids. Each experiment attempts to minimize grounding noise voltage.

7.1   Purposes of a Grounding System

Grounding noise voltage may cause failures of an electronic system. In analog systems, the lack of common reference for signal return can cause error in voltage measurements, which can result in invalid data. Similarly, in digital systems a grounding noise voltage in the DC power supply of a transistor may appear as an input of a gate and cause its false operation. Additional benefits of a good grounding system include minimizing hazards to personnel and protection of equipment.

7.1.1　Grounding of signals to minimize error voltages and number of measurements

Consider an electrical circuit as shown on Figure 7.1. The transfer function, $V(out)/V(in)$, of such a circuit can be conveniently calculated if both $V(out)$ and $V(in)$ are referred to the same reference node. If $V(out)$ uses a reference node that differs with the reference node of $V(in)$ then the difference in voltage between the two reference nodes must be known to accurately compute $V(out)/V(in)$. Specifically, if $V(in)$ uses reference node x with voltage $v(x)$ and $V(out)$ uses reference node y with voltage $v(y)$, then the error in voltages of the two reference nodes is $e = v(x) - v(y)$. The error in the transfer function, when two reference nodes are used, is therefore proportional to $v(x)/[v(x) - e]$. Using a single reference node, for both $V(out)$ and $V(in)$, not only reduces the number of measurements but also simplifies calculations. That is, using the single reference node will not require measuring $v(x)$ or $v(y)$.

The earth, with its well-designed grounding subsystem, is the ideal reference node. Its impedance is much smaller than any part of a typical electronic system. A #4 AWG power cable may have tens of hundreds of inductive reactance along its entire length. In contrast, the impedance of an earth grounding subsystem is usually limited to a few ohms.

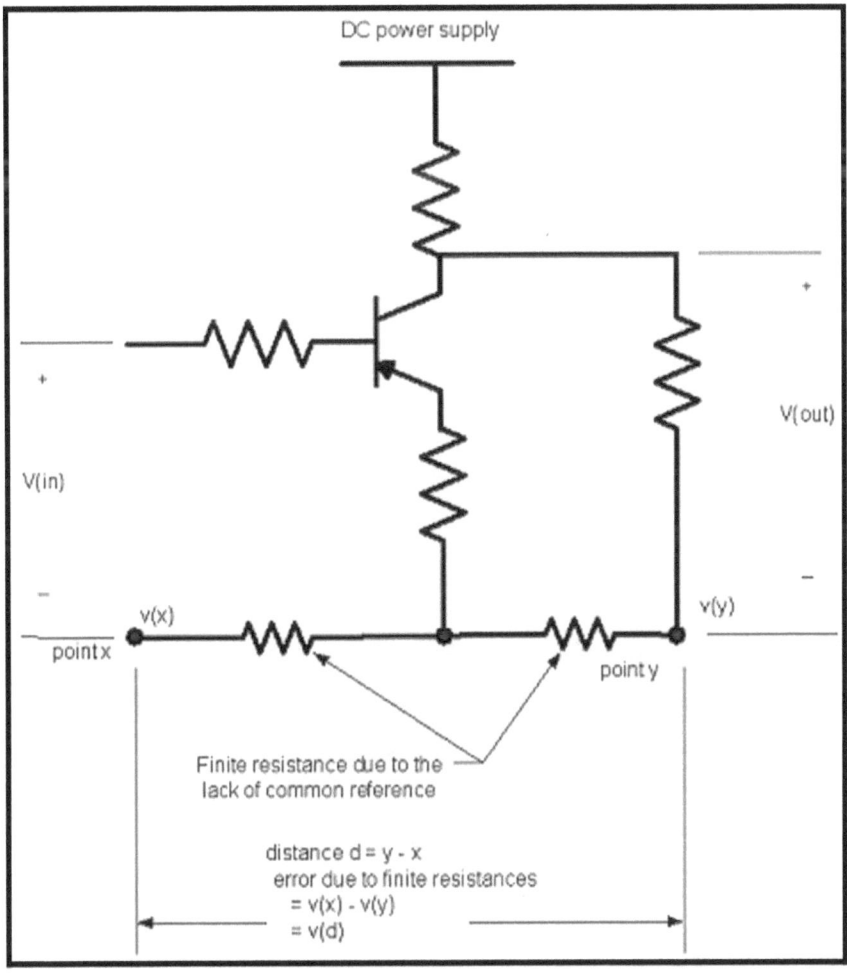

DC power supply

V(in)

V(out)

v(x)

point x

v(y)

point y

Finite resistance due to the
lack of common reference

distance d = y - x
error due to finite resistances
= v(x) - v(y)
= v(d)

**Figure 7.1 An Electronic Circuit with Different References
for Input and Output**

7.1.2   Grounding to minimize electrical hazards

Electronic circuits are usually enclosed in metallic enclosures. The
enclosure confines the radiated power of electronic parts within the
enclosure itself. Metallic enclosures are grounded to prevent electrical
hazards. If a person, for example, accidentally touches the enclosure,

by current division, the current flow in his body is proportional to the resistance of the equipment grounding conductor.

7.1.3   Protection of equipment

A grounded metallic enclosure is also a measure against an electromagnetic pulse such as lightning. Grounding essentially diverts the energy of such a pulse as quickly as possible to the earth. Without grounding, the energy may propagate to equipment and cause failures.

7.2   The Single Point Grounding System

Most electrical and electronic circuits use a return conductor from the load to the source. In a power system, such a conductor is the neutral conductor. The National Electrical Code calls this as the "grounded" conductor. Electronic systems usually designate signal names such as "DC return", "analog return", "RF return" and so on. In such systems, the return conductor must be grounded at one and only one point. Again, the basis is to minimize the error voltage as described in the previous section.

A power system has a "grounding" conductor (usually green). Do not confuse grounding conductor with grounded conductor. The main purpose of grounding conductor is for safety. Grounded conductor is for signal return.

In a power system or electronic system, the grounding conductor may be connected to as many metallic points as desired. The grounded conductor or signal return, however, must be connected to a single point only. Electronic systems with analog, digital, power, or RF signal returns must have their own separate grounded points. That is, all analog return conductors must be connected to the same analog ground. Similarly, all digital signals must be grounded to the same digital point. And so on. These points are then grounded to a single point of the *grounding* system. The net result is a star-like configuration of grounded and grounding system. At the center is a point that belongs to the *grounding* system. The rest of the points belong to the *grounded* system.

7.3 Experiments on Grounding Systems

The experiments use a rack of sensitive electronic equipment as shown on Figure 7.2. Several materials such as wires, ferrite beads, capacitors, and flat braids were used in the experiments. Results of the experiments were carefully evaluated to find the material and its configuration that yields the minimum grounding noise voltage.

7.3.1 Formulas for inductance of a ferrite bead and a circular loop

To understand the behavior of grounding noise voltage relative to materials such as a ferrite bead or a circular loop, some formulas for their inductance must be known.

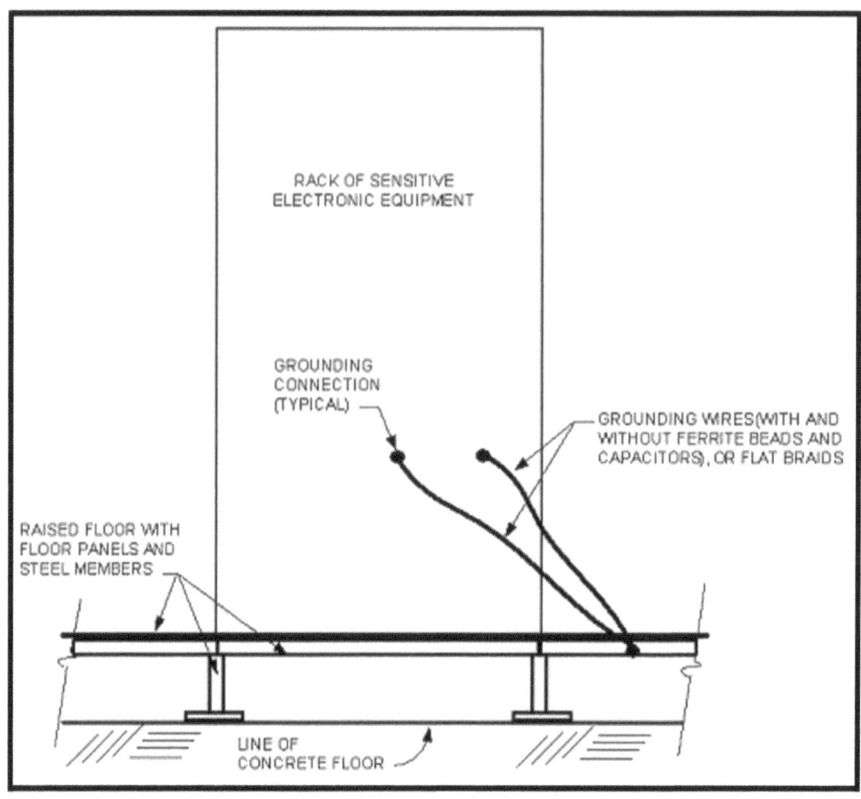

RACK OF SENSITIVE
ELECTRONIC EQUIPMENT

GROUNDING
CONNECTION
(TYPICAL)

GROUNDING WIRES(WITH AND
WITHOUT FERRITE BEADS AND
CAPACITORS), OR FLAT BRAIDS

RAISED FLOOR WITH
FLOOR PANELS AND
STEEL MEMBERS

LINE OF
CONCRETE FLOOR

**Figure 7.2 Elevation View of Enclosure Used for
Grounding Experiments**

(1)　　Wire wound around a straight length of a permeable core

Consider a wire wound around a length of permeable core. The inductance of the wire is given by

$$L = \frac{\mu_0 \mu_r N^2 A}{l} = \frac{N \Phi}{i}$$

where

μ_0 = permeability of free space,

μ_r = the relative permeability of the core,

N = number of turns,

A = cross sectional area,

l = length of the core,

Φ = flux, and

i = current.

The equation shows that lower relative permeabilty, number of turns, and cross sectional area will decrease the inductance of the wire. Additionally, longer length of the core will also decrease the inductance. Of the four variables above, the number of turns is the most critical because it is within an experimenter's control. The permeability, area, and length are controlled by its manufacturer.

(2)   Inductance of a wire wound around a ferrite bead

When a wire is wound around a ferrite bead, the resulting configuration is called a toroid. A ferrite bead is a circular core. The formula for the inductance of a toroid can be approximated from the formula for a straight core by replacing the length, l, of the core by the perimeter, $2\pi r$, of the ferrite bead.

(3)   Inductance of a wire in a circular loop

Consider next a single circular loop. The inductance of the loop is given by

$$L = r\mu_0\mu_r\left(\ln\frac{8r}{a} - 2\right)$$

where

  r = radius of the loop, and

  a = radius of the conductor.

The equation shows that lower relative permeability and radius of the loop will decrease the inductance. Also, larger cross section of the wire will decrease the inductance. Of the three variables the radius of the loop and the cross section of the wire are within an experimenter's control.

7.3.2   Cancellation of magnetic fields in a permeable core

Consider a conductor with current flowing forward. With the observer at initial end of the wire, the magnetic field in such a wire is rotating clockwise. Next, make a loop by bending the wire. The observer now sees a magnetic field that moves downward inside the loop. Outside the loop, the magnetic field is moving upward.

Next, assume two windings made of several loops. For simplicity, let the loops be in the same line from left to right. For the fields to cancel, the field inside the left coil must oppose the field inside the right coil. This means that the first loop of each coil must start at approximately the same position and their last loop must be at the opposite ends.

The two coils above uses air as the core. A permeable core, such as a ferrite bead may be used. In some of the following experiments, the magnetic fields in a core were deliberately cancelled to minimize grounding noise voltage.

7.3.3 The experiments on ferrite beads and parallel wires

Figures 7.3(a) through 7.3(h) show the results of experiments involving a low frequency ferrite bead and parallel jumpers. In general, a ferrite bead increases the inductance and therefore the voltage across the chassis and the facility steel. Additionally, Figure 7.3(c) is equivalent to Figure 7.3(g), and Figure 7.3(e) is equivalent to Figure 7.3(h).

Compare, however, Figure 7.3(b) and Figure 7.3(f). Both have one wire each. Figure 7.3(f), however, was wound to cancel the magnetic flux along the ferrite bead. As a result, its inductance is lower.

Two experiments were performed on two circular loops. The first, as shown on Figure 7.3(g), is a loop with an infinitely large radius. In this figure, the equipment grounding conductor (from the power panel) forms the other leg of the loop. Figure 7.3(h) has smaller radius. As predicted by the formula on the inductance (of a circular loop), Figure 7.3(h) must have lower inductance than Figure 7.3(g) because the former has the smaller radius.

Based on the results of the experiments, Figure 7.3(h) offers the best candidate for reducing grounding noise voltage.

7.3.4 Experiments with capacitors

A capacitor was also used in an attempt to lower the grounding noise voltage. In one case, the capacitor was placed in parallel with a conductor. The result shows that two the two parallel conductors are better than the one with a conductor in parallel with a capacitor.

In another case, the capacitor was placed in series with the combination of the ferrite bead. Again, the noise voltage did not decrease. Essentially, a capacitor allows high frequency signals and blocks low frequency signals. As a result, the series combination blocks the low frequency signals thereby increasing the low frequency noise. While the high frequency signals passes thru the capacitor, the ferrite bead blocks the same. In the end, the series combination did not lower the grounding noise voltage.

Essentially, the capacitor, whether in series or parallel with a conductor, decreases the equivalent capacitive reactance as seen from the chassis to the facility steel. The decrease increases the equivalent impedance across the chassis and facility steel.

7.3.5   The flat braid experiments

A flat braid consists of interwoven filaments of conductors. Flat braid has low inductance because of its magnetic field canceling property. Figure 7.4 shows an individual flat braid and a group of flat braids.

To see how a braid can effectively cancel magnetic fields, make two loops with the shape of an ampersand, &. Call the loops as the top loop and the bottom loop. Tracking the insides of each loop shows that the magnetic field of the top loop is downward while the magnetic field of the bottom loop is upward. In effect, the magnetic fields of two adjacent loops cancel each resulting in a lower inductance.

Tables 7.1 and 7.2 show the results of experiments with 4-#10 AWG grounding conductor, two flat braids, and four flat braids. The tables show that two flat braids are better than four flat braids. It implies that an optimum value of inductance must exist in a grounding system. In the next chapter, such a value will be used in designing a grounding system.

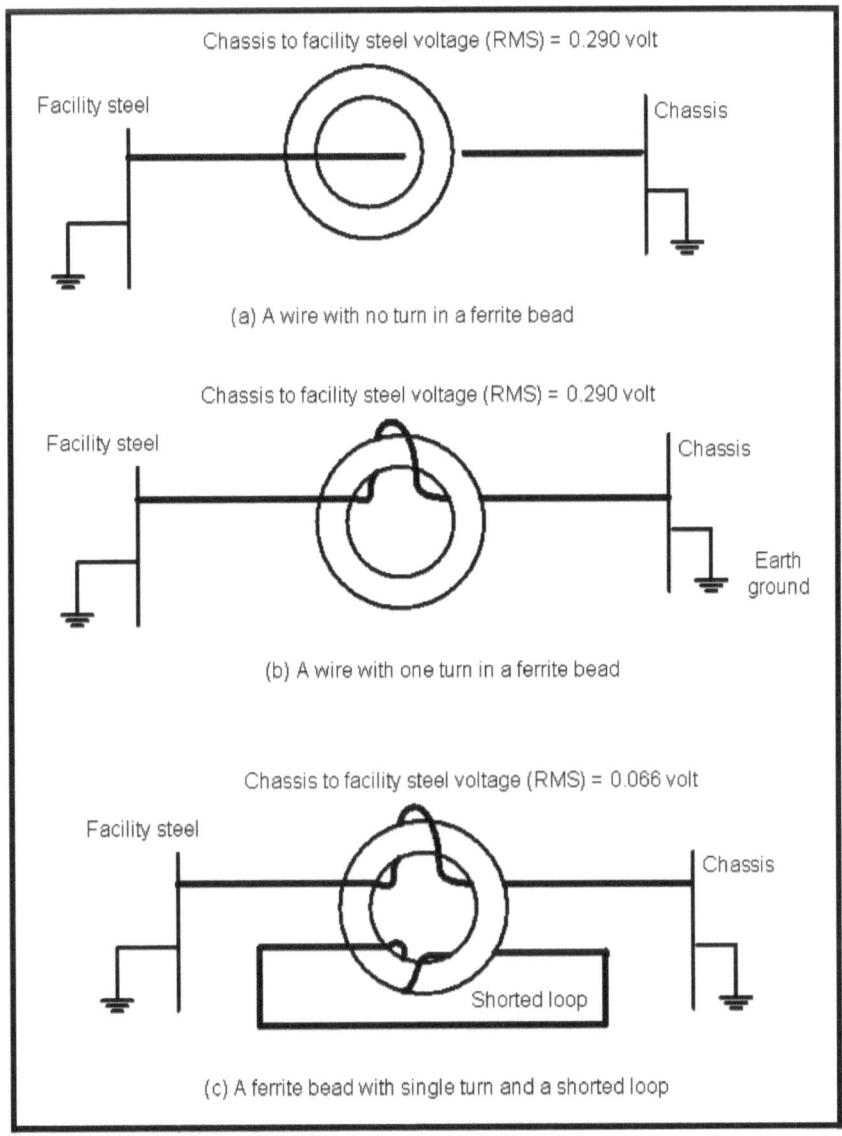

(a) A wire with no turn in a ferrite bead

(b) A wire with one turn in a ferrite bead

(c) A ferrite bead with single turn and a shorted loop

Figure 7.3 Experiments on Ferrite Bead and Parallel Wires

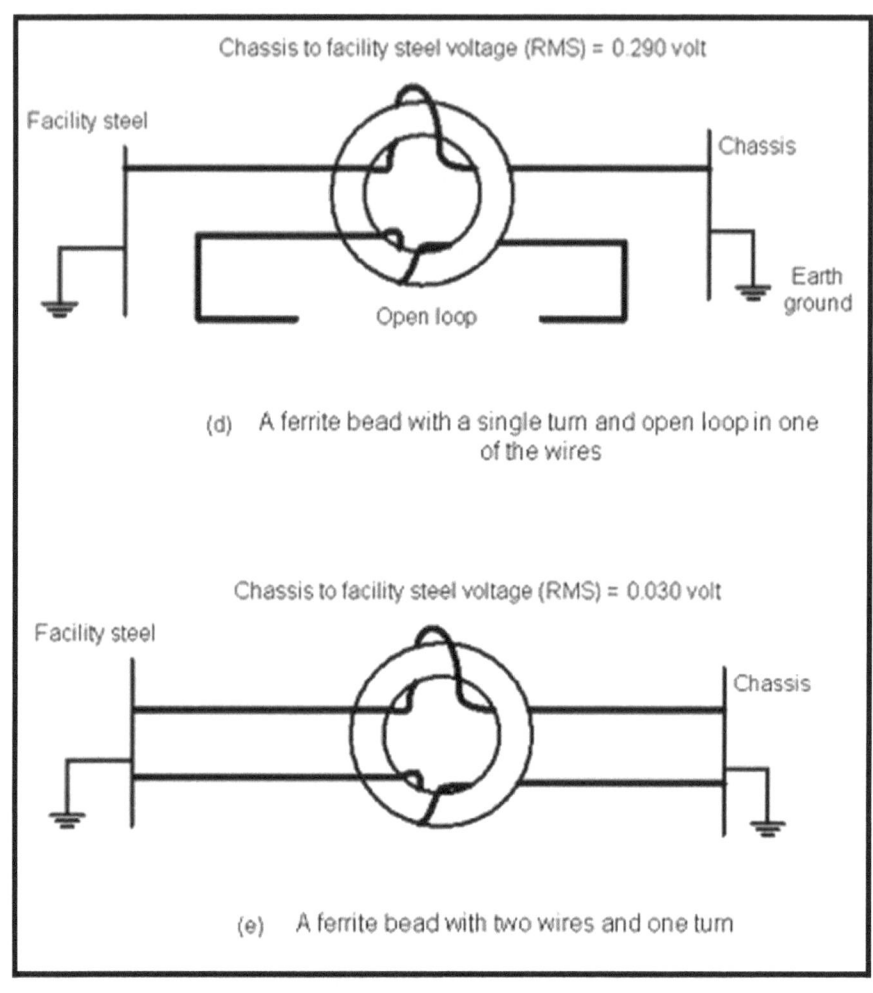

Figure 7.3 (continued) Experiments on Ferrite Bead and Parallel Wires

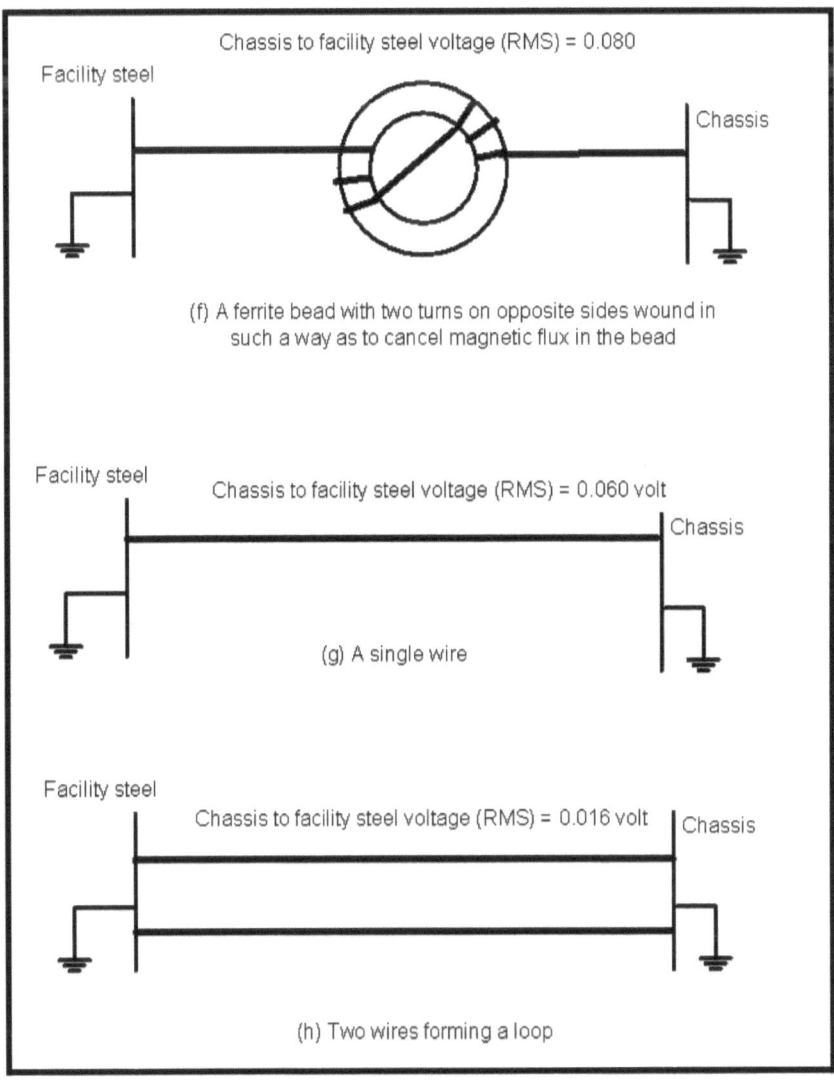

Chassis to facility steel voltage (RMS) = 0.080

Facility steel

Chassis

(f) A ferrite bead with two turns on opposite sides wound in
such a way as to cancel magnetic flux in the bead

Facility steel

Chassis to facility steel voltage (RMS) = 0.060 volt

Chassis

(g) A single wire

Facility steel

Chassis to facility steel voltage (RMS) = 0.016 volt   Chassis

(h) Two wires forming a loop

**Figure 7.3 (continued) Experiments on Ferrite Bead and
Parallel Wires**

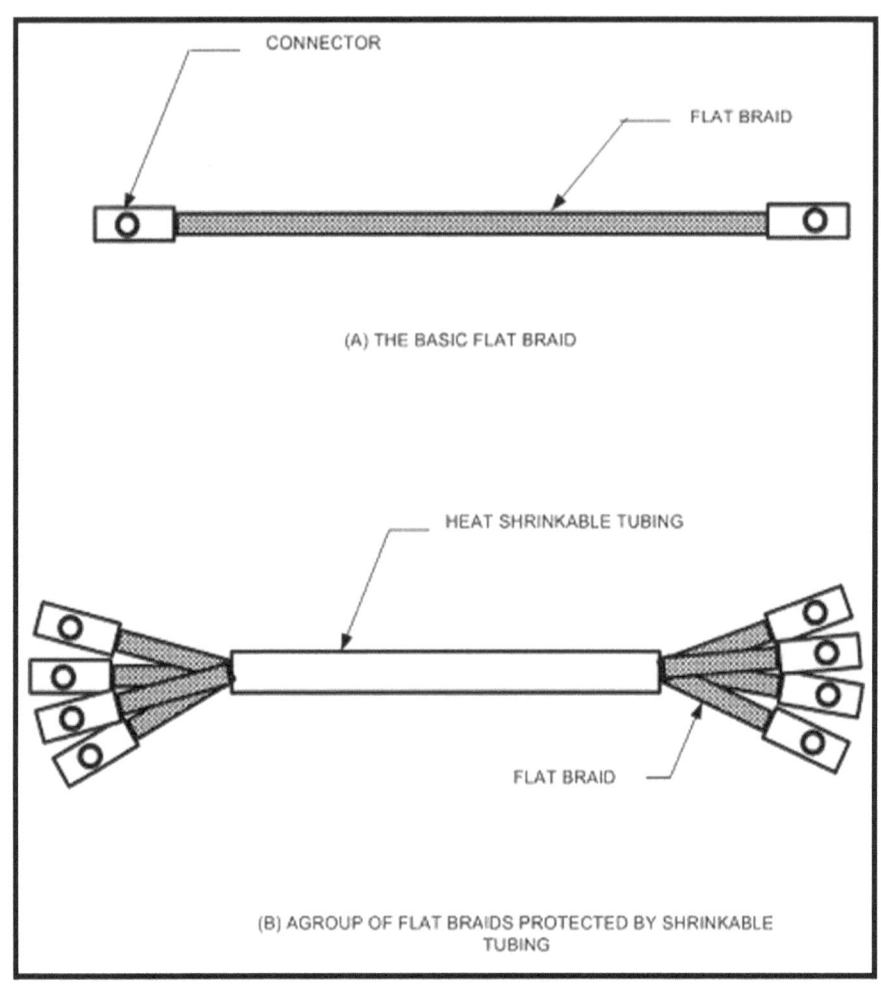

CONNECTOR

FLAT BRAID

(A) THE BASIC FLAT BRAID

HEAT SHRINKABLE TUBING

FLAT BRAID

(B) AGROUP OF FLAT BRAIDS PROTECTED BY SHRINKABLE
TUBING

Figure 7.4 A Single Flat Braid and a Group of Flat Braids

Table 7.1 Results of Grounding Experiments Using #10 AWG and Flat Braids (Using Root Mean Square Voltage)

Root mean square voltage (millivolt)	With no jumper (number of times)	With 4-#10 AWG in parallel (number of times)	With 2 flat braids (number of times)	With 4 flat braids (number of times)
0-200	6	6	9	7
200-400	7	7	9	9
400-600	2	3	2	1
600-800	2	3		2
800-1000	1			1
1000-1200	1	1		
1200-1400				
1400-1600	1			

Table 7.2 Results of Grounding Experiments Using #10 AWG and Flat Braids (Using Peak-to-Peak Voltage)

Peak-to-peak voltage (volt)	With no jumper (number of times)	With 4-#10 AWG in parallel (number of times)	With 2 flat braids (number of times)	With 4 flat braids (number of times)
0-1				
1-2	4	7	17	14
2-3	4	10	3	6
3-4	5	3		
4-5	1			
5-6	2			
6-7	1			
7-8	3			

CHAPTER 8

Recommended Procedure in
Designing a Grounding System

This chapter shows a recommended procedure on how to design the grounding system of a data center. A data center poses grounding problems not present in a typical industrial or commercial facility. That is, a data center consists of computers that are sensitive to grounding noise voltage.

Some data centers have steel raised floor with 2 feet by 2 feet floor panels. The raised floor serves as duct for airflow from a cooling unit to computers. By grounding the steel raised floor, it will be shown that grounding noise voltage can be reduced to the minimum.

The procedure consists of three parts. They are (1) theoretical proofs, (2) experiments and specific calculations, and (3) design sketches. Laplace transform was used in deriving the theoretical proofs. The proof provides the framework on how measurements have to be performed to obtain information such as frequency and attenuation of the grounding noise voltage. From the information, specific calculations show how the grounding noise power can be minimized. Finally, the calculations provide the bases for the design.

8.1   Simplified Theoretical Proofs

The simplified theoretical proofs use grounding noise voltage as the input signal. This signal is fed to a simplified circuit consisting of resistors and inductors that represent the flat braids and the #1 AWG grounding conductors. Figure 8.1 shows the details. Its transfer function depends on these circuit parameters. Examination of the transfer function shows that to maximize the -3 dB cutoff frequency, the inductance of the flat braid must be as small as possible. Similarly, avoiding resonance (of the grounding noise voltage) requires minimizing the inductance of the #1 AWG grounding conductors.

8.1.1   Proof that the flat braids must have
         loops and minimum inductance

Simplify the proof by first neglecting the inductance of the #1 AWG grounding conductors. Assume the self-impedance of the flat braids is $r + sL$, and the resistance of the #1 AWG copper is R. The ratio of the output and input voltages, and the limiting case when the -3 dB cutoff frequency is as large are given by the following equations:

$$\frac{V_o}{V_i}(s) = \frac{R}{r + R + sL}$$

$$\frac{V_o}{V_i}(s) = \frac{R/L}{(r + R)/L + s}$$

$$\lim_{L \to 0}(r + R)/L = \infty.$$

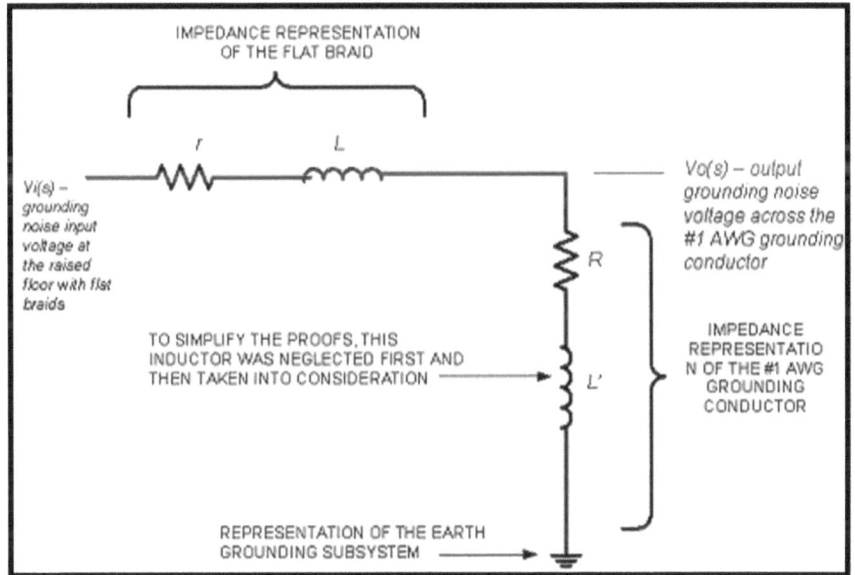

Figure 8.1 Circuit Representation of the Flat Braids and the Equipment Grounding Conductors

The -3dB cutoff frequency, which is $(r + R)/L$, approaches infinity when L is very small. Minimum L requires loops in the flat braids.

8.1.2   Proof that two or more #1 AWG grounding conductors are required

Include the effect of the inductance of the #1 AWG copper conductors. Assume that a #1 AWG copper conductor has inductance L'. Then,

$$\frac{V_o}{V_i}(s) = \frac{R+sL'}{r+R+s(L+L')}.$$

A zero at R/L' and a pole exist. Concentrate on the zero since it may cause unwanted amplification or resonance. Force the zero to positive infinity by making L' as small as possible. Doing so will minimize the risk of resonance.

That is,

$$\lim_{L' \to 0} R / L' = \infty$$

Again, L' may be minimized by using two or more #1 AWG conductors in parallel. Note that two parallel conductors form a loop.

8.1.3   Check for stability of section 8.1.2 for possible resonance at the noise frequency

Assume that the wavelength of the noise is much greater than the physical length of the grounding conductor. In such a case, the grounding conductor is inductive. The effect of the capacitance between the wire and the ground plane or any neighboring wire may be neglected.

In general, stability analysis requires estimating the phase angles at the zeros of the numerator and the poles of the denominator. Next, the phase margin of the design is evaluated as the difference of the phase angle at 0 dB and 180 degrees. As long as the absolute value of this difference is at least 45 degrees but less than 180 degrees then the circuit is stable. Feedback, in the form of other external noise, could exist in the circuit.

(1)   Phase calculations at $s = 0$ (DC)

The phase of the numerator at $s = 0$ is zero degree (no imaginary part). Similarly, the denominator has a phase of zero degree also.

(2)   Phase calculations at the zero (of the numerator)

Since L' is forced to zero, the ratio of the imaginary part and real part is close to zero. Hence, the phase angle at zero (numerator) is almost equal to zero. Similarly, the inductance, L, of the flat braid is almost equal to zero. Hence, the phase at the pole of the denominator is almost zero. The difference in the two phase angles is, therefore, almost zero.

(3)   Phase calculation at the pole (of the denominator)

At the pole frequency, the phase of the numerator is almost zero. However, the phase of the denominator is 45 degrees. The sum of the phases is almost 45 degrees. Consider the worst case when the 0 dB gain is at the pole. Then the worst-case phase is 45 degrees.

(4)   Phase margin calculations (assuming feedback exists in the system)

The phase margin is the absolute value of the difference between 180 degrees less 45 degrees (from case (3) above). Therefore, the phase margin is 135 degrees. Since the phase margin is greater than 45-degrees then the design is stable.

8.1.4   Estimate of the further reduction in noise

The ratio of the new and the old impedances (without consideration of zeros and poles), from the raised floor to the earth, provides a measure of further reduction in noise. This ratio is estimated to be about 0.25 corresponding to a reduction of -12.0 dB. By making the frequency of the pole and zero as large as possible, additional reduction of at least – 6 dB is possible. Therefore, the total reduction in noise is about -18 dB or less.

8.2   Measurements and Specific Calculations

With the theoretical proofs of the design established, the next step is to perform measurements of the grounding system to find the frequencies with the maximum noise level. The technique described in the previous chapter may be used in the measurements.

8.2.1   Measurements

It is common to see different noise levels at two different frequencies. For example, noise level at 2 MHz maybe three times the noise level at 100 KHz. This means that the grounding system produces resonance at 2 MHz. In the following section, the frequency of 2 MHz will be used as the example of the design frequency.

8.2.2   Specific calculations comparing the inductive reactance of copper and the existing steel grid at the noise frequency

Recall that the purpose of Figure 8.1 is to replace the existing grounding system of a steel raised floor by one with flat braids and copper grounding conductors. To compare how better the proposed design is compared to the existing raised floor, specific calculations of the inductive reactance of copper and steel are required. The idea is to use the material with the least inductance and reactance.

The following formulas for inductance and inductive reactance were from [32]. Units in the formula require the metric or SI units.

$$L = \frac{\mu_0 \mu_r}{2\pi} \cosh^{-1}(h/a) \text{ (in Henry/meter)}$$

where

  h = height of a wire over a ground plane, and
  a = radius of the wire.

The inductive reactance is

$$|X_L| = |j2\pi fL| = 2\pi fL \text{ (in ohms/meter)}$$

Outline of the calculations follows.

(1)   Inductive reactance of a single grounding conductor (#1 AWG) over a ground plane – includes permeability as a parameter

  •   Height, h, of the wire over the ground plane = 0.01 meter
  •   Radius, a, of the wire = 0.0042 meter (NEC 2005 Table 8, Conductor Properties)
  •   Relative permeability of the medium = 1.25 (copper)
  •   Noise frequency in Hz = 2.0 MHz (function of switching frequencies in the power supplies)
  •   Inductive reactance = 4.75 ohms per meter.

(2)   Inductive reactance of existing steel grid – similar to the above conditions except the conductor is made of steel

  •   Relative permeability of the steel = 100.0
  •   Noise frequency = 2.0 MHz
  •   Inductive reactance = 345.75 ohm/meter. This is worse than

copper above. Additionally, the steel grid has discontinuities at each cell making the inductance much larger.

(3)   Conclusion. The copper grounding wire is better than the steel by about 72.8 times (minimum). Use copper wires for grounding.

(4)   Division of Noise Power between the Copper Wires and the Steel Grid

- Ratio of the noise power that will flow in the copper grounding wires:

$$N_g = \frac{345.75}{4.75 + 345.75} = 0.98 \text{ (or 98\%)}$$

- Ninety-eight percent will flow in the copper wires and two percent in the steel grid.

(5)   Maximum length of flat braid against RF resonance (IEEE Standard 142-1991)

- The maximum length of the flat braid, from equipment to the grid, to reduce RF resonance at 2.0 MHz is 7.5 meters.

Figure 8.2 shows the flow of noise in the raised floor under two conditions: (1) without flat braids and grounding conductors; and, (2) with flat braids and grounding conductors. That is, with the flat braids and grounding conductors installed, the noise in the existing raised floor is only 2% of the noise when there is no flat braid and grounding conductor.

8.3 Sketches of Plans and Details

The above calculations show that the flat braids and the equipment grounding conductors must have loops. Loops in the flat braids may be synthesized by installing them in a grid pattern. Paralleling equipment grounding conductors is equivalent to having loops with them. Figure 8.3 shows the plan of the grid on the raised floor. Its installation detail is shown on Figure 8.4. Figure 8.5 shows the schematic diagram of the interconnection between the grounding grid system and the earth grounding subsystem. The earth grounding subsystem consists of grounding rods spaced no more than the length of a rod.

To sample 2 MHz grounding noise signals, install no more than 7.5 meters of grounding conductor from equipment to the raised floor.

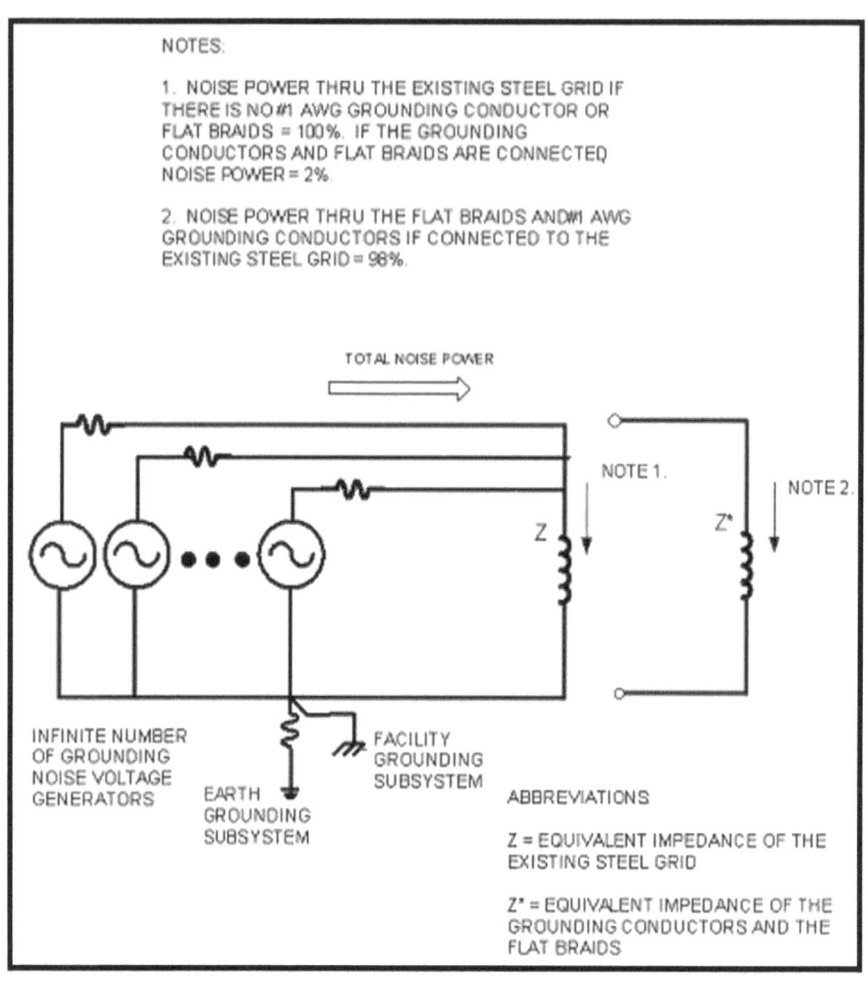

**Figure 8.2 The Flow of Noise in the Steel Grid of a
Raised Floor with and without Flat Braids
and Equipment Grounding Conductors**

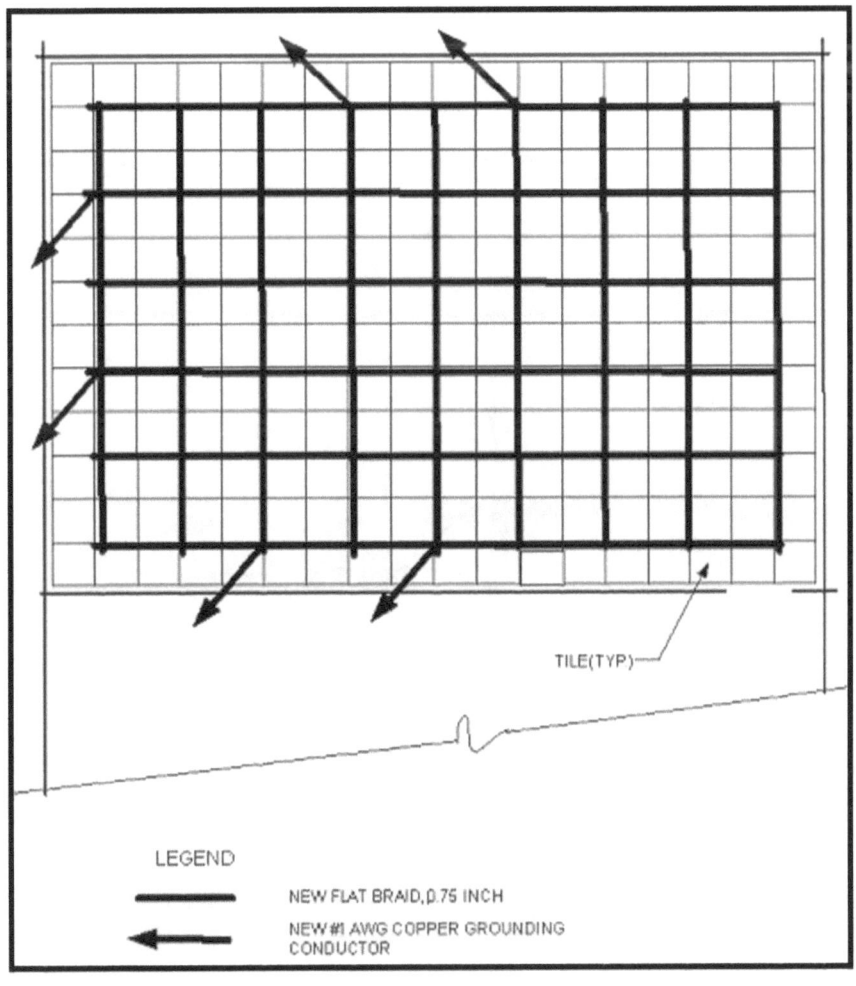

TILE(TYP)

LEGEND

NEW FLAT BRAID, 0.75 INCH

NEW #1 AWG COPPER GROUNDING
CONDUCTOR

**Figure 8.3 Plan of the Raised Floor with New Flat Braids
and Grounding Conductors**

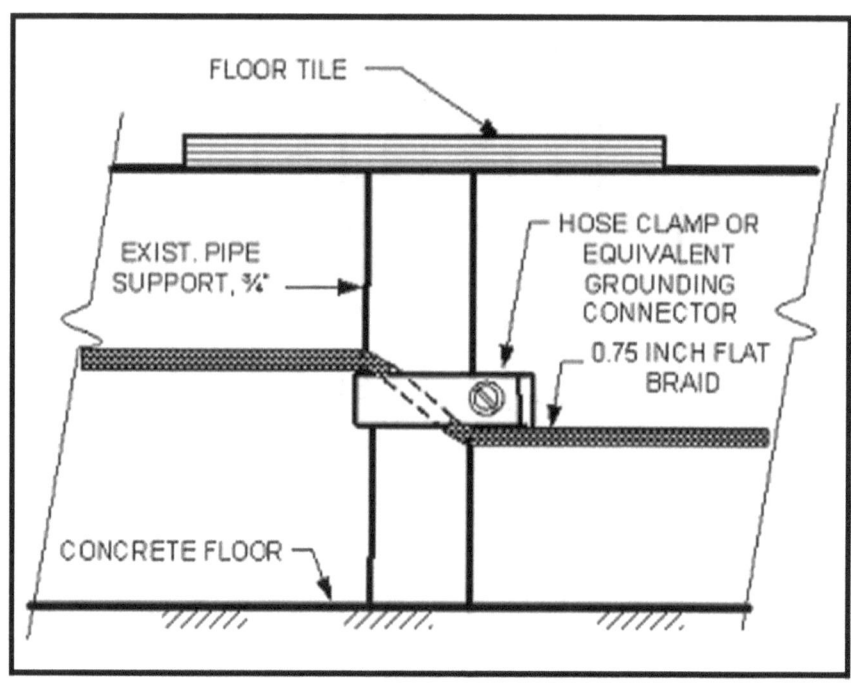

Figure 8.4 Installation Details of the Flat Braids under the Raised Floor (Side View)

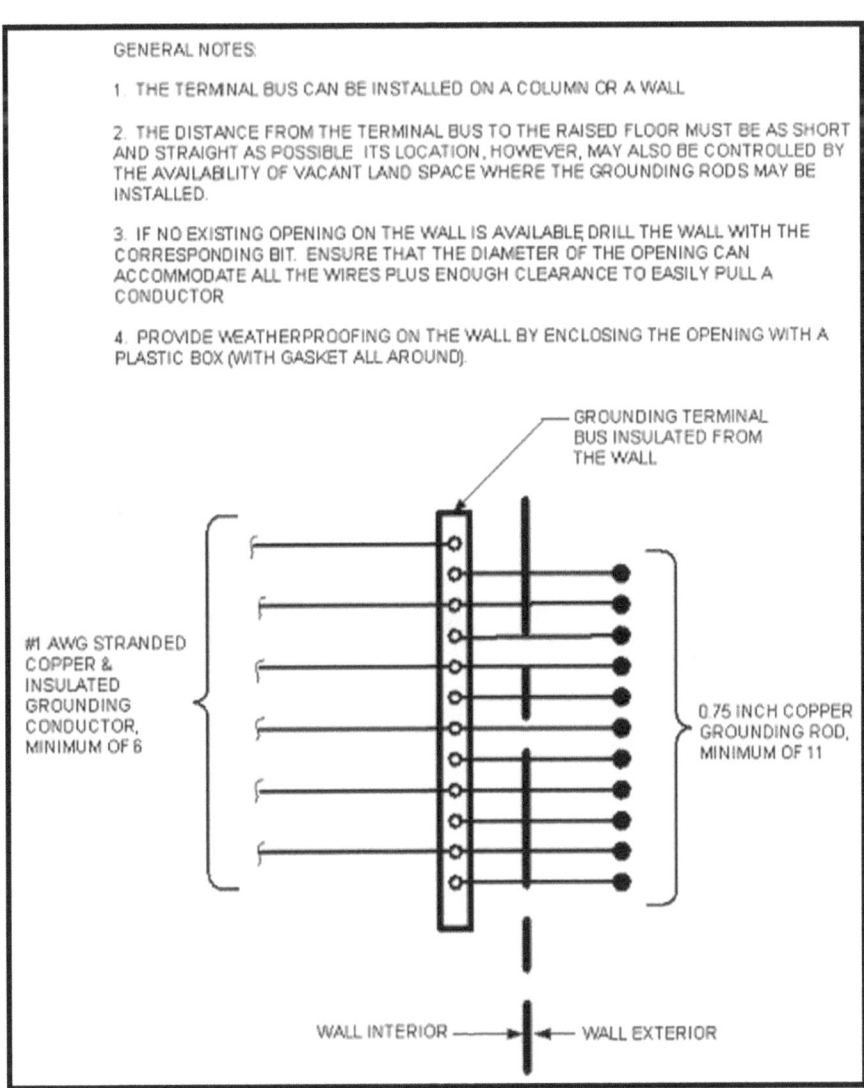

GENERAL NOTES:

1. THE TERMINAL BUS CAN BE INSTALLED ON A COLUMN OR A WALL.

2. THE DISTANCE FROM THE TERMINAL BUS TO THE RAISED FLOOR MUST BE AS SHORT AND STRAIGHT AS POSSIBLE. ITS LOCATION, HOWEVER, MAY ALSO BE CONTROLLED BY THE AVAILABILITY OF VACANT LAND SPACE WHERE THE GROUNDING RODS MAY BE INSTALLED.

3. IF NO EXISTING OPENING ON THE WALL IS AVAILABLE, DRILL THE WALL WITH THE CORRESPONDING BIT. ENSURE THAT THE DIAMETER OF THE OPENING CAN ACCOMMODATE ALL THE WIRES PLUS ENOUGH CLEARANCE TO EASILY PULL A CONDUCTOR.

4. PROVIDE WEATHERPROOFING ON THE WALL BY ENCLOSING THE OPENING WITH A PLASTIC BOX (WITH GASKET ALL AROUND).

GROUNDING TERMINAL BUS INSULATED FROM THE WALL

#1 AWG STRANDED COPPER & INSULATED GROUNDING CONDUCTOR, MINIMUM OF 6

0.75 INCH COPPER GROUNDING ROD, MINIMUM OF 11

WALL INTERIOR ——▶ ◀—— WALL EXTERIOR

Figure 8.5   The Use of the Grounding Terminal Bus to Interconnect Grounding Conductors from the Raised Floor to the Earth Grounding Subsystem

115

Appendix A

Transformation of Shunt Resistance and Capacitance to Equivalent Series Resistance and Inductance

```
// import the required libraries from Java

import java.io.*; // the default input/output library in Java
import java.util.*; // the default library of utilities in Java

// start of the class

public class ParallelToSerialTransformation {

    //start of the method

    public static void getTransformation() throws IOException {

    Scanner keyboard = new Scanner(System.in);
    FileWriter f = new FileWriter("Results.txt", true); //text file for
        saving results
    PrintWriter pWriter = new PrintWriter(f);
    String M;
    String m0;
```

// prompt the user for the shunt resistance

```
m0 = "\n\n\t > INPUT the resistance (in ohms): ";
System.out.print(m0);
double R = keyboard.nextDouble();
M = m0 + R;
pWriter.print(M);
```

// prompt the user for the shunt capacitance

```
m0 = "\n\n\t > INPUT the capacitance (in Farad): ";
System.out.print(m0);
double C = keyboard.nextDouble();
M = m0 + C;
pWriter.print(M);
```

// prompt the user for the frequency of signal

```
m0 = "\n\n\t > INPUT the frequency (in Hertz): ";
System.out.print(m0);
double dFreq = keyboard.nextDouble();
M = m0 + dFreq;
pWriter.print(M);
```

// the following are the calculations

```
double d1Term = 2.0 * Math.PI * dFreq; // frequency in radians
     per second
d1Term = d1Term * R * C;
double d1Factor = d1Term * R;
d1Term = d1Term * d1Term;
```

```java
double d2Factor = 1.0 + d1Term;
d2Factor = 1.0 / d2Factor;

double Req = d2Factor * R;
double Leq = -d1Factor * d2Factor;

// result for the equivalent resistance in series

m0 = "\n\n\t ~ Equivalent resistance (ohm) = ";
M = m0 + Req;
System.out.print(M);
pWriter.print(M);

// result for the equivalent inductance in series

m0 = "\n\n\t ~ Equivalent inductance in series with the resistance
    (Henry) = ";
M = m0 + Leq;
System.out.print(M);
pWriter.print(M);

//close the text file

f.close();

    }//end of the method
}// end of the class
```

Java is a trademark of Sun Microsystems, Inc.

References

Materials in this book are based on industry references, handbooks, textbooks, and other publications. They are grouped accordingly as shown on the following list:

References on Industry Practices

1. IEEE Recommended Practice for Applying Low-Voltage Circuit Breakers Used in Industrial and Commercial Power Systems (IEEE Blue Book, IEEE Std 1015-1997), The Institute of Electrical and Electronics Engineers, Inc., New York, NY, 1997, pp. 1-152.

2. IEEE Recommended Practice for Electric Power Systems in Commercial Buildings (IEEE Gray Book, IEEE Std 242-1990), The Institute of Electrical and Electronics Engineers, Inc., New York, NY, 1991, 13- 49, 397-497.

3. IEEE Recommended Practice for Grounding of Industrial and Commercial Power Systems (IEEE Green Book, IEEE Std 142-1991), The Institute of Electrical and Electronics Engineers, Inc., New York, NY, 1992, pp.13- 49, 145-163.

4. IEEE Recommended Practice for Protection and Coordination of Industrial and Commercial Power Systems (IEEE Buff Book, IEEE Std 242-2001), The Institute of Electrical and Electronics Engineers, Inc., New York, NY, pp. 2001, 11- 336.

5. NEC 2005, NFPA 70: National Electrical Code, National Fire Protection Association, Inc., Quincy, MA, 2004. pp. 70-26 to 70-153, 70-516 to 70-518, 70-635 to 70-637.

Recommended Handbooks on Electrical Engineering

6. Chen, Wai-Kai, Editor-in chief, The Circuits and Filters Handbook, Massachusetts, CRC Press, Inc., 1995.

7. Reeve, Whittman, D., Subscriber Loop Signaling and Transmission Handbook: Digital, The Institute of Electrical and Electronics Engineers, Inc., New York, NY, 1995.

8. Whitaker, Jerry C., Editor-in-chief, The Electronics Handbook, Massachusetts, CRC Press, Inc., 1995.

References on Mathematics, Complex Numbers, and Partial Differential Equation

9. Abramowitz, Milton, and Stegun, Irene A., editors, Handbook of Mathematical Functions, 9th printing, New York: Dover Publications, 1965.

10. Kreyzig, Erwin, Advanced Engineering Mathematics, New York: John Wiley & Sons, Inc., 1979.

References on the Propagation Constant of a Transmission Line

11. Carson, Ralph S., High-frequency amplifiers, New York: John Wiley & Sons, Inc., 1982.

12. Gonzalez, Guillermo, Microwave Transistor Amplifiers: Analysis and Design, New Jersey: Prentice-Hall, 1996.

13. Hayt, William, H., Engineering Electromagnetics, fourth edition, New York: McGraw-Hill, 1981.

14. Liao, Samuel Y., Microwave Circuit Analysis and Amplifier Design, New Jersey: Prentice-Hall, 1987.

15. Sayre, Cotter, W., The Complete RF Technician's Handbook, 2nd edition, Indiana: Howard W. Sams & Company, 1998.

References on the use of Laplace Transform in the Analysis and Design of Electronic Circuits

16. Carson, Chen, Active Filter Design, New Jersey: Hayden Book Company, 1982.

17. D'Azzo, John J, and Houpis, Constantine, H, Linear Control System Analysis and Design, 2nd edition, New York: McGraw-Hill, 1981.

18. Glasford, Glen M., Analog Electronic Circuits, New Jersey: Prentice-Hall, 1986.

19. Hayt, William H, and Kemmerly, Jack E., Engineering Circuit Analysis, 3rd edition, New York: McGraw-Hill, Inc., 1978.

20. Raven, Francis, H, Automatic Control Engineering, New York: McGraw-Hill, 1961, 64-88.

21. Stanley, William D., Operational Amplifiers with Linear Integrated Circuits, New York: Macmillan, 1989.

22. Van Valkenberg, M. E., Network Analysis, 3rd edition, New Jersey: Prentice-Hall, 1974.

References on Modulation Theory and its Application on the Design of Systems such as the DC Power Supply

23. Evans, Alvis J, Basic Digital Electronics, Illinois: Master Publishing, Inc., 1986.

24. Freer, John, Computer Communications and Network, 2nd edition, New Jersey: IEEE Press, 1996.

25. Lathi, B. P., Modern Digital and Analog Communication Systems, New York: Holt, Rinehart, and Winston, 1983.

26. Sinnema, William, Digital, Analog, and Data Communication, Virginia: Reston Publishing Company, Inc., 1982.

27. Wozencraft, John M, and Jacobs, Irwin M., Principles of Communication Engineering, Wiley & Sons, Inc., 1965.

References on Bulk 60-Hz Power Generation, Transmission, and Wiring Systems

28. Gilbert, Elliott, M., and Nellist, John G., Telecommunications Wiring for Commercial Buildings, New York: IEEE Press, 1996.

29. Grainger, John J. and Stevenson, William D, Power System Analysis, New York: McGraw-Hill, 1994.

30. Siskind, Charles, S., Electrical Control Systems in Industry, New York: McGraw-Hill, 1963.

31. Stevenson, W.D., Jr., Elements of Power System Analysis, New York: McGraw-Hill, 1982.

References cited from the Internet

32. Technick.net, Utilities, Inductance Calculator, (accessed June 2, 2008).

33. Wikipedia contributors, "Bode plot," Wikipedia, The Free Encyclopedia, http://en.wikipedia.org/w/index.php?title=Bode_plot&oldid=215557591 (accessed May 29, 2008).

34. Wikipedia contributors, "Laplace transform," Wikipedia, The Free Encyclopedia, http://en.wikipedia.org/w/index.php?title=Laplace_transform&oldid=215593324 (accessed May 29, 2008).

35. Wikipedia contributors, "Transfer function," Wikipedia, The Free Encyclopedia, http://en.wikipedia.org/w/index.php?title=Transfer_function&oldid=198193742 (accessed May 29, 2008).

Index

3-phase, 5-wire power distribution system 13
4-wire and 5-wire 3-phase systems 1
120-volt, single-phase 3
208-volt, 3-phase, 4-wire 2, 4
208-volt, single-phase 4
600-volt class 17, 44

A

admittance 34, 40
analog return 90, 91
attenuation constant 34, 40, 56

B

Balanced Loads 9
boundary 64, 73
branch breakers 5

C

capacitance 18, 19, 44, 45, 78, 106, 118
characteristics impedance 40, 41, 45
characterization of a signal 87
circuit analysis 55
circuit breakers 17
circular loop 91, 93, 95
complex conjugate 42
complex power 48, 54
conductance 34, 40, 41
convolution theorem 55, 61, 75

D

data center. 103
dB 75, 77, 78, 82, 104, 106, 107

DC power supply 63, 73, 74, 75, 88
delay 10, 33, 45, 48, 73, 81
DeMoivre's formula 39, 42

E

electrical fault 17
electromagnetic pulse 90
equipment grounding conductor 1, 2, 3, 4, 6, 8, 9, 10, 11, 12, 13, 14, 90, 95
equipment grounding terminal bar 5
error 88, 90

F

Fault calculations 17
ferrite beads 87, 91, 95
field effect transistor 76
filters 75, 76
flat braid 96, 104, 107, 110
flat braids 87, 91, 96, 101, 104, 105, 108, 110, 111
frequency response 75, 76, 77, 80
fully redundant system 47, 54, 56

G

"grounded" conductor 90
grounding conductor 3, 6, 8, 9, 14, 90, 96, 106, 109, 110, 111
grounding noise voltage 1, 87, 88, 91, 94, 95, 96, 103, 104

H

harmonics 14, 81
hazards 88, 89
high frequency signals 96

T

thermal limit 42
three-phase load 2, 7, 8
three phase power distribution system 2
tracing a fault 63, 64, 74, 87
transfer function 55, 56, 61, 62, 88, 104
transmission line 33, 34, 36, 42, 54, 55, 56, 57
transmitter and receiver of noise 14
trigonometric identities 48, 54
tripping mechanisms 17

U

unbalanced current 5, 8, 11, 12
unbalanced currents 1, 13

V

voltage drop 7, 82
voltage regulation 8, 42
volt-ampere 8

W

wavelength 40, 41, 45, 106
wye secondary winding 2, 5

Z

zeros 55, 106, 107

www.ingramcontent.com/pod-product-compliance
Lightning Source LLC
Chambersburg PA
CBHW030800180526
45163CB00003B/1106